AIRO Springer Series

Volume 6

The AIRO Springer Series focuses on the relevance of operations research (OR) in the scientific world and in real life applications.

The series publishes peer-reviewed only works, such as contributed volumes, lectures notes, and monographs in English language resulting from workshops, conferences, courses, schools, seminars, and research activities carried out by AIRO, Associazione Italiana di Ricerca Operativa - Optimization and Decision Sciences: http://www.airo.org/index.php/it/.

The books in the series will discuss recent results and analyze new trends focusing on the following areas: Optimization and Operation Research, including Continuous, Discrete and Network Optimization, and related industrial and territorial applications. Interdisciplinary contributions, showing a fruitful collaboration of scientists with researchers from other fields to address complex applications, are welcome.

The series is aimed at providing useful reference material to students, academic and industrial researchers at an international level.

Should an author wish to submit a manuscript, please note that this can be done by directly contacting the series Editorial Board, which is in charge of the peer-review process.

THE SERIES IS INDEXED IN SCOPUS

More information about this series at http://www.springer.com/series/15947

Adriano Masone • Veronica Dal Sasso •
Valentina Morandi
Editors

Optimization and Data Science: Trends and Applications

5th AIROYoung Workshop and AIRO PhD School 2021 Joint Event

AIRO
ASSOCIAZIONE ITALIANA DI RICERCA OPERATIVA
OPTIMIZATION AND DECISION SCIENCE

Springer

Editors
Adriano Masone
Department of Electrical Engineering
and Information Technology
University of Naples "Federico II"
Naples, Italy

Veronica Dal Sasso
Optrail
Rome, Italy

Valentina Morandi
Faculty of Science and Technology
Free University of Bozen
Bolzano, Italy

ISSN 2523-7047 ISSN 2523-7055 (electronic)
AIRO Springer Series
ISBN 978-3-030-86285-5 ISBN 978-3-030-86286-2 (eBook)
https://doi.org/10.1007/978-3-030-86286-2

This Springer imprint is published by the registered company Springer Nature Switzerland AG.
The registered company address is: Gewerbestrasse 11, 6330 Cham, Switzerland

Preface

This book contains the proceedings of the 5th AIRO Young Workshop and AIRO PhD School 2021 joint event on "Optimization and Data Science: Trends and Applications," held online, from February 8 to 12, 2021.

This volume presents methodological and application-oriented contributions relating to optimization and data science methods, as well as a large variety of applications in computer science, healthcare, logistics, and transportation. The 14 accepted contributions are organized in 5 topical parts: Data Science and Machine Learning, Healthcare, Logistics, Optimization for Control Systems, and OR in Industry. In each part, the contributions are listed alphabetically by the last name of the first author.

In the first part, "Data Science and Machine Learning", the reader will find the following contributions.

- **Reinforcement learning for the Knapsack Problem, by Pierotti et al.** The complexity of combinatorial optimization (CO) problems makes it difficult to find the optimal solution via an exact solution method. In recent years, machine learning (ML) has brought immense benefits in many research areas, including heuristic solution methods for CO problems. Among ML methods, reinforcement learning (RL) seems to be the most promising method. In this work, the authors investigate an RL framework to achieve solutions for the knapsack problem. The presented algorithm finds close to optimal solutions for instances up to one hundred items, which leads to the conjecture that RL and self-attention may be major building blocks for future state-of-the-art heuristics for other CO problems.
- **Potential sales estimates of a new store, by Tozzi and Guarino.** This work describes a real application consisting in the estimation of new point of sales (PoS) potential. The potential of a PoS is measured in terms of estimated sales in its second year of activity with respect to the opening date of the contract. The authors propose an original approach based on the combined use of gradient boosting and convolutional neural network. The aim is to support sales managers with an automatic tool returning the most promising PoS.

- **Sells optimization through product rotation, by Tozzi and Guarino.** A company aims at maximizing the vending machines (VMs) sales within a points of sale (PoS) network. The sales of a VM decrease proportionally to the number of days the VM remains in the same PoS. A "novelty effect" arises when the VM is moved in a different PoS. The "novelty effect" leads to an increase of the VM sales in the first period of exposure of the VM in the different PoS. In this work, the authors optimally solve the problem of determining the rotation of the VMs within the PoS network which maximizes the VM sales.

In the second part, "Healthcare", the reader will find the following contributions.

- **Gathering avoiding centralized pedestrian advice framework: an application for COVID-19 outbreak restrictions, by Dal Sasso and Morandi.** Due to the COVID-19 pandemic, maintaining a safe distance among pedestrians becomes crucial in big pedestrian networks. Looking at personal goals, such as walking through the shortest path, could lead to congestion phenomena on both roads and crossroads, violating the imposed regulations. To this end, the authors suggest a centralized multi-objective approach able to assign alternative fair paths for users while maintaining the congestion level as low as possible.
- **A MILP formulation for the reorganization of the blood supply chain in Italian regions, by Mancuso et al.** Blood is a vital resource for a human being and it is crucial for surgeries and medical emergencies. Thus, blood supply chain management has generated great interest in terms of efficient and effective policy making and system design. In this context, a mixed-integer linear programming formulation is presented to determine the optimal location and the number of blood facilities on a regional scale, with the aim of minimizing system costs while guaranteeing a good standard service level.

In the third part, "Logistics", the reader will find the following contributions.

- **Instance generation framework for green vehicle routing, by Andrade and Usberti.** In the green vehicle routing problems (G-VRP), electric vehicles with limited autonomy can recharge at alternative fuel stations (AFSs) to keep visiting customers. To the best of the authors' knowledge, the G-VRP scientific literature accounts for only two sets of instances. Hence, in this chapter they propose a framework for generating relevant sets of instances for G-VRP, based on solving a maximum leaf spanning tree problem to address the location of AFSs. Two G-VRP variants are considered, where consecutive AFSs visits are allowed, and where they are not allowed.
- **An optimization model for service requests management in a 5G network architecture, by Colajanni and Sciacca.** The authors present a three-tier supply chain network model consisting of a fleet of UAVs organized as a FANET (fly ad hoc network) connected to each other with direct wireless links, managed by a fleet of UAV controllers, whose purpose is to provide 5G network slices on demand to users and devices on the ground. The aim of this contribution is to determine the optimal distributions of request flows. The authors formulate a

constrained optimization problem and derive the associated variational inequality formulation. A numerical example is performed to validate the effectiveness of the model.

- **A MIP model for freight consolidation in road transportation considering outsourced fleet, by Viera and Munari.** The chapter addresses the freight consolidation problem with outsourced fleet. The work is motivated by a Brazilian real-life situation of a major manufacturer of school supplies which has to optimally assign shipments to vehicles in a way to minimize the total transportation cost. The challenge is given by the complex pricing table provided by outsourcing companies. In fact, outsourcing cost functions follow a piecewise linear behavior of the cost function, which makes the consolidation more difficult. A mixed-integer linear programming (MIP) model, which fully represents the problem, is proposed, and costs reductions of more than 44%, with respect to the usual freight consolidation policy of the company, have been observed.

In the fourth part, "Optimization for Control Systems", the reader will find:

- **Energy-oriented inter-vehicle distance optimization for heterogeneous E-platoons, by Coppola et al.** Connected and autonomous vehicles (CAVs) have the potential to improve the energy efficiency of transportation systems. In particular, the inter-vehicle distance plays a critical role for energy-saving purposes. On this basis, this contribution proposes a novel optimization algorithm to compute the optimal gap distance in a heterogeneous platoon of electric CAVs by exploiting a distance-dependent air drag coefficient formulation.
- **Optimization-based assessment of initial-state opacity in Petri Nets, by De Tommasi et al.** When dealing with security and safety problems, discrete events systems could be a convenient way to model the behavior of distributed dynamical systems. In this context, a valuable property of a system is the opacity. This property is related to the capability of hiding a secret to external observers. In this chapter, leveraging the mathematical representation of Petri Nets, the authors present a feasibility problem with integer optimization variables verifying a sufficient condition which permits to assess if a system is not opaque.
- **Eco-driving adaptive cruise control via model predictive control enhanced with improved Grey Wolf optimization algorithm, by Petrillo et al.** In this chapter, a novel ecological adaptive cruise control system for an autonomous electric vehicle is suggested. The proposed system is able to drive its motion while minimizing as much as possible its energy consumption. To this aim, the authors considered a nonlinear model predictive control method enhanced with an offline computational intelligence-based optimization algorithm.

In the fifth part, OR in Industry, the reader will find:

- **Optimizing and evaluating a maintenance strategy for multi-component systems, by Bautista Bárcena and Torres Castro.** Maintenance optimization is a key challenge in engineering and industry. In this chapter, a system with monitored and non-monitored components is considered where components respectively deteriorate following a gamma process and a Poisson process.

The goal is to minimize the time slot between inspections and, hence, to find the minimum cost maintenance plan. Different techniques for the optimization process are employed such as a Monte Carlo simulation and meta-heuristic algorithms.

- **Metal additive manufacturing: Nesting vs. scheduling, by Kucukkoc.** In additive manufacturing (AM), parts are produced through a layer-by-layer production process. Selective laser melting (SLM) is a popular AM technology used to build metal components. Although it may seem a high-cost process at first glance, it can be compensated with efficient planning and scheduling systems. In this chapter, the author aims to investigate the relationship between nesting and scheduling when planning and scheduling SLM machines.
- **System and methods for blockchain-inspired digital game asset management, by Ragnoni**. In this chapter, the optimal design of a fully managed ledger database providing a transparent, immutable, and cryptographically verifiable platform for managing the creation of digital assets (e.g., game ticket) and the transfer of asset's ownership between users is described. The aim of this contribution is to present the usefulness of the resulting platform, its features, and its functioning mechanism.

As editors of the volume, we thank AIRO and AIROYoung, the invited lecturers, the authors, and the researchers who spent their time for the review process, contributing to improve the quality of the selected contributions. A special thanks should be addressed to the AIROYoung representatives and the Operations Research Group of the Department of Electrical Engineering and Information Technology of the University "Federico II" of Naples who organized the joint event.

Finally, we express our gratitude to Springer for its strong support and cooperation during the event and the publishing process.

Rome, Italy Veronica Dal Sasso
Naples, Italy Adriano Masone
Bolzano, Italy Valentina Morandi

Contents

About the Editors

Adriano Masone is a Postdoctoral Researcher at the Department of Electrical Engineering and Information Technology of the University of Naples "Federico II." He obtained his PhD in Information Technology and Electrical Engineering in 2020 at the University of Naples "Federico II." In 2018–2019, he was a Visiting Scholar at the Robert H. Smith School of Business of the University of Maryland, Maryland, USA. His areas of research include exact and heuristic solution methods for complex combinatorial and network optimization problems with application to healthcare, transportation, logistics, production planning and scheduling, among others.

Veronica Dal Sasso holds a PhD in Mathematics from the University of Padova. After the completion of the PhD, she held a postdoc at Lancaster University, where she was involved in the OptiFrame project, a project funded by the European Union under the Horizon 2020 agreement. In 2018, she moved to Rome and started working at Optrail as Operations Research Scientist. Optrail is a company devoted to providing innovative decision support solutions for the railway industry. She is the founder and current treasurer of AIROYoung, and she was part of the organizing committees for the 3rd, 4th, and 5th AIROYoung Workshops.

Valentina Morandi is an Assistant Researcher in the Science and Technology faculty at Free University of Bolzano/Bozen. She obtained her PhD in Analytics for Economics and Business in 2017 at the University of Bergamo jointly with the University of Brescia. Her research activities focus on congestion avoiding traffic assignment techniques and fair models for logistics, public transportation, and pedestrians. She teaches Operations Research in the Mechanical Engineering course, and she won the Euregio Best Young Researcher in 2019 for her research project on models for fair vehicular traffic management.

Part I
Data Science and Machine Learning

Reinforcement Learning for the Knapsack Problem

Jacopo Pierotti, Maximilian Kronmueller, Javier Alonso-Mora, J. Theresia van Essen, and Wendelin Böhmer

Abstract Combinatorial optimization (CO) problems are at the heart of both practical and theoretical research. Due to their complexity, many problems cannot be solved via exact methods in reasonable time; hence, we resort to heuristic solution methods. In recent years, machine learning (ML) has brought immense benefits in many research areas, including heuristic solution methods for CO problems. Among ML methods, reinforcement learning (RL) seems to be the most promising method to find good solutions for CO problems. In this work, we investigate an RL framework, whose agent is based on self-attention, to achieve solutions for the knapsack problem, which is a CO problem. Our algorithm finds close to optimal solutions for instances up to one hundred items, which leads to conjecture that RL and self-attention may be major building blocks for future state-of-the-art heuristics for other CO problems.

Keywords Reinforcement learning · Multi-task DQN · End-to-end · Knapsack problem · Transformer · Self-attention

1 Introduction

In recent years, machine learning (ML) has shown super-human capabilities in speech recognition, language translation, image classification, etc. [4, 12, 16]. Lately, more and more combinatorial optimization (CO) problems have been studied under the lens of machine learning [3]. Among these CO problems, NP-hard problems are of interest because, so far, solving them to optimality (via so-called exact methods) takes exponential time; thus, for many classes of CO problems, obtaining good solutions for large or even medium sized instances in reasonable time can only be achieved by exploiting handcrafted heuristics. Instead of creating

J. Pierotti (✉) M. Kronmueller · J. Alonso-Mora · J. T. van Essen · W. Böhmer
TU Delft, Delft, Netherlands
e-mail: J.Pierotti@tudelft.nl; M.Kronmuller@tudelft.nl; J.AlonsoMora@tudelft.nl; J.T.vanEssen@tudelft.nl; J.W.Bohmer@tudelft.nl

A. Masone et al. (eds.), *Optimization and Data Science: Trends and Applications*, AIRO Springer Series 6, https://doi.org/10.1007/978-3-030-86286-2_1

a heuristic by hand, one can also use ML to train a neural network to predict an almost optimal solution for given or randomly generated CO instances [3]. This way heuristics can be learned without expert knowledge of the problem domain, which is also called *end-to-end training*. Reinforcement learning (RL) seems to be the most promising end-to-end method to solve combinatorial problems [2]. In fact, in difference to supervised ML, RL does not need to know the solutions to given training instances to learn a good heuristic. This way one can learn a heuristic without any domain knowledge and, in principle, one could find a heuristic that works better than any a human would be able to design. RL has been used to train the neural networks used by heuristics designed to solve CO problems [9–11], including the knapsack problem (KP) [2]. The aim of this paper is to develop an RL end-to-end algorithm for the knapsack problem based on attention [16], in difference to prior work that used either recurring neural networks (RNN) or convolutional neural networks (CNN) [4, 12] (which are popular NN for end-to-end methods). By developing such an algorithm for a relatively easy CO problem (the KP) [13], we want to assess if RL with attention can be a fruitful method to tackle other, more complex, CO problems, which will be the focus of future research.

The remainder of the paper is organized as follows. The formulation of the KP, our motivations on how and why we use attention and not RNNs or CNNs, and model architecture are presented in Sect. 2. The training distributions (i.e. benchmarks of instances) used for testing and evaluating as well as the computational results are detailed in Sect. 3. Finally, in Sect. 4, we illustrate our conclusions.

2 Problem Formulation and Background Information

The knapsack problem (KP) is one of the most studied CO problems [13]. As input, we have a set of objects (denoted by set N) and a knapsack of capacity W. Each object $i \in N$ has a positive profit p_i and a positive weight w_i. The objective of the problem is to maximize the sum of the profits of the collected objects without violating the capacity constraint. Introducing binary variables x_i, which assume value one if object $i \in N$ is selected and zero otherwise, we can write the problem as follows:

$$\max \sum_{i \in N} x_i p_i \tag{1}$$

$$\sum_{i \in N} x_i w_i \leq W \tag{2}$$

$$x_i \in \{0, 1\} \qquad \forall i \in N. \tag{3}$$

The objective function (1) maximizes the total profit of the selected objects, constraint (2) acts as the capacity constraint, and constraints (3) force the variables to be binary. This integer linear program (ILP) belongs to the class of NP-hard problems [13], which means that the computation time for obtaining optimal solutions with known exact solution methods grows exponentially with the number of objects. A simple yet very powerful heuristic is to sort the objects in non-increasing order of their ratio, i.e., $q_i = \frac{p_i}{w_i}$ for $i \in N$, and collect them in order as long as constraint (2) is respected (collecting non-consecutive objects is allowed). In the following, we refer to this as the *simple heuristic*.

2.1 Reinforcement Learning Framework

In this section, we give an overview of how we implemented our algorithm. For more details on RL, we refer the reader to [14]. Our algorithm belongs to the area of multi-task RL [17], where a task is an instance of the knapsack problem. The difference between single and multi-tasks is that: in single-task, we want to learn a policy to always solve the same (instance of a) problem; in multi-task, we want to learn a policy to solve a family of different instances of a problem (or even different problems). Moreover, while in single-task the initial state is always the same, this does not hold in multi-tasks. In order to describe a state in our case, we first define how we embed the objects into vectors. At any given time step, each object $i \in N$ is uniquely associated to a vector. Each vector is of the form $t_i = [p_i, w_i, q_i, \bar{x}_i, u]$, where $\bar{x}_i$ is a binary parameter assuming value zero when object i has already been selected or cannot be selected due to the capacity constraint (2) and one otherwise, and u is the residual capacity of the knapsack (i.e., W minus the weights of the already selected objects).

We name the selection of an object an *action*. Actions (A) are chosen based on the Q-value of each object (see Sect. 2.1.1). The algorithm that determines the Q-values is called the *agent* (see Sect. 2.2). In RL, a *state* represents the available information about the process at a given moment. We represent the observation of a state by the matrix obtained stacking all the $|N|$ object vectors together. Given a non-final state, the agent has to select an action; however, not all objects can be chosen in any state. While choosing an action, the non-selectable objects are momentarily removed, which is called masking. In our case, generic action (object) i is masked when $\bar{x}_i$ equals zero. The initial state has $u = W$ and $\bar{x}_i = 1$ for all $i \in N$, while we define a state as final if $\bar{x}_i = 0$ for all $i \in N$. Our algorithm sequentially selects objects until no additional object can be selected, in which case the algorithm terminates.

Given a state, each action leads to a new state and a reward. In our case, the reward r of choosing object $i \in N$ is the profit of the chosen object (i.e. $r = p_i$ when choosing object $i \in N$). The series of states in between an initial and a final state is called an *episode*. The final objective of an RL algorithm is to maximize the (discounted) cumulative reward observed in an episode. In general, we discount the future reward to avoid problems arising with very long or non-finishing episodes. In our case, episodes are relatively short and they always terminate (worst case

Table 1 Summary of the definitions needed for our reinforcement learning framework

Name	Definition
Task	An instance of the KP
Multi-task RL	RL algorithm to solve a family of tasks (virtually any KP problem in our case)
Action	The selection of an object
State	The available information (profits, weights, which objects have been selected and which have not,..) at a given time moment
Initial state	State where no objects have been selected yet
Final state	State where no more objects can been selected
Episode	Series of visited states from the initial state to the final state
Masking	The removal of the unselectable actions
Reward	The profit of the selected object
Transition	A sequence of an old state, a chosen action, an observed reward, and a new state
Minibatch	A set of non-consecutive[1] transitions
Training distribution	Distribution from which we draw the instances to train the algorithm

scenario, they terminate in $|N|$ steps); so, there is no need to discount the future rewards. We call the sequence of old state s, chosen action a, observed reward r, and new state s' a *transition*. A set (of fixed size, in our case) of non-consecutive[1] transitions is called a *minibatch*. The transitions in the minibatches are used to compute the loss (which is needed in order to learn) in the learning step (Sect. 2.1.2). Finally, we train and evaluate the algorithm by solving randomly drawn tasks from the so called *training distribution*. Table 1 summarizes the introduced definitions.

2.1.1 Double Q-Learning

Our RL algorithm falls under the general umbrella of Q-learning [18]. Given a state s, and a set of possible actions A, the idea of Q-learning is to estimate the expected future cumulative rewards for each possible action (called Q-values $Q(s, a), \forall a \in A$) and select one action based on an exploration/exploitation strategy. On one hand, exploration is fundamental to search the state-action space. In fact, in (non-deep) Q-learning, if one could explore for an infinite amount of time, the optimal Q-values would be retrieved. On the other hand, the agent should concentrate more on promising actions to improve convergence to an optimal policy. As exploration/exploitation strategy, we use ϵ-greedy, which greedily chooses the best action (i.e. the action with the highest Q-value) with probability 1-ϵ or a random action with probability ϵ. Often the Q-learning algorithm can be too optimistic while

[1] Transitions do not have to be consecutive, but, by chance, they could be.

estimating the Q-values. One common solution to this problem is to adopt double Q-learning [7]. In deep RL, double Q-learning is enforced by having two identically structured neural networks. The current network is used to select the best action at the next state while the other one (called the *target* network) is used to compute the Q-value of the next state. In this work, we use a similar method which helps stabilizing our results. The difference being that the Q-values are always computed via the current network and the target network is used to determine the action. Naming Q' the function to compute the Q-values associated with the target network, our revised Bellman equation becomes (see also Sect. 2.1.2):

$$Q(s, a) = r(s, a) + Q(s', \arg\max_a(Q'(s', a))). \tag{4}$$

Equation (4) is needed in the learning step (see Sect. 2.1.2), where the parameters of the Q function are tuned in such a way that the distance between $Q(s, a)$ and $r(s, a) + Q'(s', \arg\max_a(Q(s', a)))$ is minimized.

2.1.2 Learning Strategy

Our algorithm works by generating and solving new tasks of different dimensions (i.e. $|N|$ is not a constant between two different tasks). Let us assume that we train our algorithm to solve instances of k different sizes. Every time a new instance is generated and solved, all the transitions are stored in a replay buffer [20]. Our algorithm has k different fixed-size replay buffers (one for each possible dimension of $|N|$) where transitions are stored with a FIFO (first in first out) strategy. A FIFO policy guarantees that the algorithm always keeps in memory the newest generated information. Transitions of instances with the same dimensions are stored in the same buffer. When a minibatch is needed, we randomly choose one of the k replay buffers and extract a minibatch from there. Given the multiple replay buffers, each state in the minibatch has the same dimension and, thus, can be stack together, easing the computation. It is important to note that different tasks have different gradient magnitude: a task with 100 objects is likely to have a different gradient than a task with 2 objects. In fact, we are using a NN to approximate the Q-values and, reasonably, the approximation becomes more and more difficult (thus less and less accurate) with an increasing number of objects. A less accurate Q-value approximation would likely lead to greater gradient magnitudes; thus, different tasks present different gradient magnitude. However, since each time we choose the replay buffer uniformly at random, we are averaging the gradients; thus, we are not introducing any bias. When the algorithm has accumulated *enough* transitions in the replay buffers, it begins to learn. We do so by selecting, uniformly at random, from one random replay buffer, t transitions (or all the transitions if less than t transitions are present in that replay buffer). Transitions which have never been selected before have priority over transitions that were. We call these t transitions a *minibatch*. For generic transition i (s_i, a_i, r_i, s_i') in the minibatch, we compute the loss as:

$$loss_i = \Big(Q(s_i, a_i) - \big(r_i + Q(s_i', \arg\max_a (Q'(s_i', a))\big)\Big)^2 \qquad (5)$$

Then, we backpropagate the average of the t losses. Sometimes, the loss function is so steep that blindly following its gradient would lead outside of the region where the gradient is meaningful. To prevent this, we clip the gradient [19] to a maximum length of 0.1. The parameters of the agent are updated via the RMSprop method[2] [5]. Finally, the *target* network is updated via a soft-update [6], i.e., naming p any generic parameter of the agent, p_t its corresponding one in the target network and τ (constant equal to 0.05 in our case) the soft-update parameter: $p_t \leftarrow (1-\tau)p_t + \tau p$.

2.2 The Agent

The agent receives the observations of the states and outputs the Q-values. It is composed by three main blocks, all using ReLU as activation function. The first and last block are composed of two fully connected linear layers of dimension 512 each. The first block enlarges the feature space of each object vector from five to 512 and the last block reduces the features to one (the Q-value). The second block is a transformer (Sect. 2.2.1). In most CO problems, there is no clear ordered object structure. Even if we introduce an arbitrary order, the problem would be permutation invariant. In the KP, a permutation of the elements would neither change the optimal solution of the problem nor its structure. For this reason, we decide to base our agent on self-attention, which is permutation invariant (unlike CNNs or RNNs). While most agents for end-to-end approaches involve CNNs and/or RNNs [11], we conjecture that, for the KP and other CO problems, the effectiveness of an algorithm does not lie within those structures. In fact, CNNs are an excellent tool to extract local features [12], but they are only useful when there is a clear ordered object structure (such as pixels in an image). RNNs sequentially embed a sequence of inputs, where each output depends also on the sequence of previous inputs. This is very useful when states are partially observable [4]; however, the KP satisfies the Markov property, i.e., the distribution of future states depends only on the current state. This memoryless property makes the problem Markovian. Thus, given the Markovian property of our problem and the absence of an underlying ordered structure, we decide to base our implementation on a variation of the transformer [16] without CNNs or RNNs. The transformer accepts as input a variable-length (d_t) tuple of objects (where all objects have the same dimension d_o) and returns a tuple of same length and dimension (d_o for each single output, d_t for the whole

[2] http://www.cs.toronto.edu/~tijmen/csc321/slides/lecture_slides_lec6.pdf.

tuple). It is composed by a series of multi-head attention mechanisms in a layer structure (see Sect. 2.2.1). Attention is a powerful mechanism that allows to look at the input and generate a context vector based on how much each part of the input is relevant for the output. Doing so, the algorithm learns to isolate from a set of features the one(s) relevant for that particular state.

2.2.1 Self-Attention, Multi-Head, and Multi-Layer Transformer

Self-attention is a powerful ML technique that takes a set of objects and returns an equally sized set of vectors. In our case, the objects taken as input are matrices, called queries Q, keys K, and values V, which are three different linear transformations of the object vectors of size d_q, d_n, and d_n, respectively. Self-attention is a function that measures the similarity of queries and keys with a dot product; then, a softmax of that similarity is used to weight the values in a linear combination. So, naming W^Q, W^K, and W^V the matrices of learnable parameters for the linear transformations and $\bar{S} \in \mathbb{R}^{n \times d_n}$ the embedded state observation (i.e., the matrix obtained by stacking the object vectors of dimension 512), we obtain:

$$\text{Attention}(Q, K, V) = \text{softmax}\left(\frac{QK^\top}{\sqrt{d_n}}\right) V = \text{softmax}\left(\frac{(\bar{S}W^Q)(\bar{S}W^K)^\top}{\sqrt{d_n}}\right)(\bar{S}W^V).$$

Instead of a single self-attention mechanism on vectors of dimension d_n, [16] discovered that it was beneficial to linearly project the queries, keys, and values h times (called a *head*; hence, multi-head) with different, learned linear projections on a smaller dimension of size $d_v = \dfrac{d_n}{h}$. The outputs are computed in parallel, concatenated, and reprojected once again (via a learnable matrix W^0). Formally, this becomes:

$$\text{MultiHead}(\bar{S}) = [head_1, \cdots, head_h]W^0,$$

where $head_i = \text{Attention}(\bar{S}W_i^Q, \bar{S}W_i^K, \bar{S}W_i^V)$ and W_i^Q, W_i^K, W_i^V are all learnable matrices for all $i \in [1, \ldots, h]$. This multi-head self-attention mechanism is repeated for L layers. Each layer is composed of two units which both produce outputs of the same dimension as their input, i.e., d_n. The first unit is indeed the multi-head self-attention mechanism, the second unit is a fully connected feed-forward network with ReLUs. Both these units adopt also a residual connection and a layer normalization [1]. The residual connection was proven to facilitate learning [8].

2.3 Model Architecture

In each instance, all objects are normalized such that the maximum profit and weight is one. The agent has a two layer fully connected neural network to expand the 5 features of a vector into 512 features. The resulting vector is fed to a transformer encoder[3] with six layers and eight heads per layer. Normalization is applied after each layer. After the transformer, another two fully connected neural network layers are used to reduce the 512 features to a single one (the Q-value associated with the action of selecting the corresponding object). The learning rate of the optimizer is set to 10^{-6} and ϵ linearly decreases with the episode number from one to 0.05. Each replay buffer can store up to a maximum of 10^5 transitions, the minibatch size is set to 512, and the soft update parameter τ is set to 0.05. The overall structure of the algorithm is given in Algorithm 1. For a total of 10^5 times, the algorithm generates and solves one instance. Its transitions are saved in the replay buffer and the algorithm takes a learning step. In order to partially fill the replay buffers, the algorithm starts to learn only after the 512th iteration. During the training, ten equally spaced greedy test evaluations over one hundred randomly generated instances are conducted in order to assess the algorithm progress.

Algorithm 1: RL algorithm overview

for $i = 0, \cdots, 10^5$ **do**
    task $\leftarrow$ generate new task
    transitions $\leftarrow$ solve the task with an ϵ-greedy policy
    store transitions in replay buffer
    if $i \geq 512$ **then**
        learning step
    if $i \mod 10^4 = 0$ **then**
        evaluate the algorithm with a greedy policy

3 Computational Results

Two different training distributions are used to generate the tasks. In the first distribution, $|N|$ is chosen uniformly at random between 2 and 100 every time a new instance is generated. Moreover, the profit and weight of each object are also chosen uniformly at random in the closed interval $[10^{-6}, 1]$. A lower bound of 10^{-6} is enforced to avoid numerical errors. We call this distribution *random*.

[3] For details see https://pytorch.org/docs/stable/generated/torch.nn.TransformerEncoder.html. Many optional parameters were set to the default values, such as the feedforward dimension was set to 512 and the probability of dropout to 0.1.

The second distribution (named *Pisinger*) are some of the *small, large*, and *hard* instances taken from [13]. These Pisinger instances were generated in order to be difficult to be solved via a MILP solver. These small, large, and hard instances are further subdivided in six, six, and five groups, respectively. From these groups, we select instances with 20, 50, and 100 objects. Each pair group-number of objects contains one hundred instances, for a total of 3200 instances (because not all groups have the 20 objects instances).

We train our algorithm twice from scratch, thus obtaining two different versions of the same model. We train the first version exclusively on the random instances while we train the second one exclusively on the Pisinger instances. We evaluate the trained algorithms both on random instances and on Pisinger's. When evaluating and testing, we compare our results with the simple heuristic (see Sect. 2) which achieves, on average, 99% of the optimal solution's value (hence, it is a good measure for comparison). In Figs. 1 and 2, every result is normalized with respect to

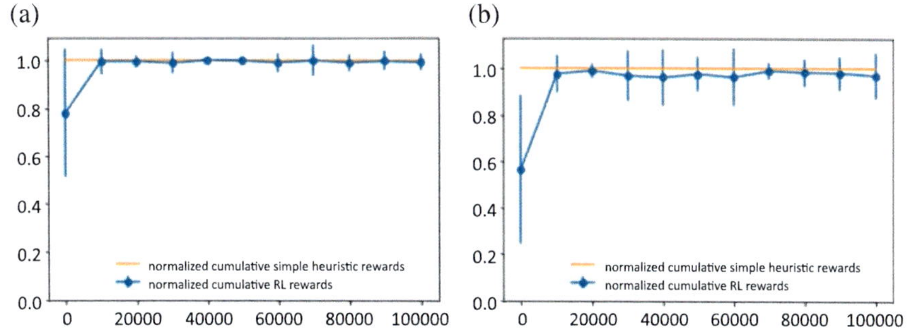

Fig. 1 During training evaluation on 100 random instances. On the x-axis, the number of iterations and on the y-axis, the averaged normalized cumulative reward are shown. The blue dots indicate the average cumulative reward, the vertical lines indicate the standard deviation. (**a**) Training on the random distribution. (**b**) Training on the Pisinger distribution

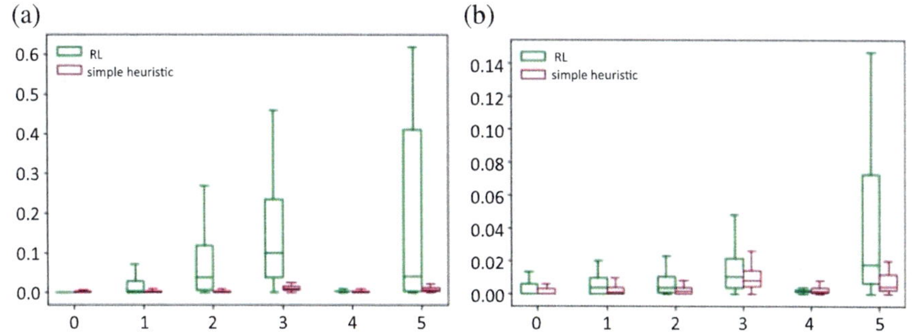

Fig. 2 Evaluation on the hard, 100 objects Pisinger instances. Green lines for the RL and purple lines for the simple heuristic. On the x-axis, different groups of instances, on the y-axis, the gaps to optimality are displayed. Please note the different scale of the y-axis. (**a**) Training on random distribution. (**b**) Training on Pisinger distribution

the optimal solutions (in the Pisinger distribution) or with respect to the heuristic solution. Figure 1 displays the evaluations of the algorithm during training on one hundred random instances. For the sake of brevity, we report only the most meaningful results, i.e., the *hard* Pisinger instances with one hundred objects. Figure 2a shows the boxplot of the gap to the optimal solution for the *hard* Pisinger instances of the algorithm trained on the random distribution. Although the results are overall satisfactory, the algorithm trained on random instances performs badly on some types of Pisinger instances. The most likely reason is that the algorithm trained on random instances has an extremely small probability of seeing some Pisinger instances (which have been handcrafted), thus it does not generalize over those particularly complex instances. On the other hand, when the algorithm is evaluated on randomly generated instances (Fig. 1a), results are very close to the heuristic solution, thus, to the optimal solution. Figure 2b displays the same gap for the algorithm trained on the Pisinger distribution. In this case, results are very satisfactory since the algorithm consistently achieves near-optimal solutions. Also while evaluating on randomly generated instances (Fig. 1b), results are very close to the heuristic solution, thus to the optimal solution; however, results are slightly worse than the results obtained by the algorithm trained on the random distribution. As expected, we conclude that training the algorithm on randomly generated instances boosts performance in the average case, but it is less effective to complex instances, while training the algorithm on the Pisinger distribution performs (slightly) worse on the average case, but is much more robust (both on the random and on the complex Pisinger instances).

4 Conclusion

In this work, we introduced a deep Q-learning framework with a transformer as the main deep architecture for the KP problem. The algorithm achieves results very close to optimality within a split second on instances up to one hundred objects. These results are promising; however, in the KP, also a simple conventional heuristic returns very solid results. Nonetheless, our results suggests that "attention is all you need" may also hold in end-to-end methods for CO problems. Future research will explore a transformer-based RL method on other CO problems where conventional heuristics fail to give good solutions in a short amount of time. Moreover, our algorithm computes the Q-values which are difficult quantities to estimate. Instead, one could aim to learn directly the policy with whom to take actions (i.e. the probability distribution of the actions for a given state). This policy could be learned by firstly using behavioural cloning [15] (to imitate the heuristic), and secondly RL, to explore more possible state-action combinations.

Acknowledgments This research was supported in part by Ahold Delhaize. All content represents the opinion of the author(s), which is not necessarily shared or endorsed by their respective employers and/or sponsors.

References

1. Ba, J.L., Kiros, J.R., Hinton, G.E.: Layer normalization. Preprint (2016). arXiv:1607.06450
2. Bello, I., Pham, H., Le, Q.V., Norouzi, M., Bengio, S.: Neural combinatorial optimization with reinforcement learning. Preprint (2016). arXiv:1611.09940
3. Bengio, Y., Lodi, A., Prouvost, A.: Machine learning for combinatorial optimization: a methodological tour d'horizon. Eur. J. Oper. Res. **290**, 405–421 (2021)
4. Bontemps, L., McDermott, J., Le-Khac, N.-A.: Collective anomaly detection based on long short-term memory recurrent neural networks. In: International Conference on Future Data and Security Engineering, pp. 141–152. Springer (2016)
5. Duchi, J., Hazan, E., Singer, Y.: Adaptive subgradient methods for online learning and stochastic optimization. J. Mach. Learn. Res. **12**(7), (2011)
6. Fox, R., Pakman, A., Tishby, N.: Taming the noise in reinforcement learning via soft updates. Preprint (2015). arXiv:1512.08562
7. Hasselt, H.: Double q-learning. Adv. Neural Inf. Process. Syst. **23**, 2613–2621 (2010)
8. He, K., Zhang, X., Ren, S., Sun, J.: Deep residual learning for image recognition. In: Proceedings of the IEEE Conference on Computer Vision and Pattern Recognition, pp. 770–778 (2016)
9. Joshi, C.K., Cappart, Q., Rousseau, L.-M., Laurent, T., Bresson, X.: Learning TSP requires rethinking generalization. Preprint (2020). arXiv:2006.07054
10. La Maire, B.F., Mladenov, V.M.: Comparison of neural networks for solving the travelling salesman problem. In: 11th Symposium on Neural Network Applications in Electrical Engineering, pp. 21–24. IEEE (2012)
11. Nazari, M., Oroojlooy, A., Snyder, L.V., Takáč, M.: Reinforcement learning for solving the vehicle routing problem. Preprint (2018). arXiv:1802.04240
12. O'Shea, K., Nash, R.: An introduction to convolutional neural networks. Preprint (2015). arXiv:1511.08458
13. Pisinger, D.: Where are the hard knapsack problems? Comput. Oper. Res. **32**(9), 2271–2284 (2005)
14. Sutton, R.S., Barto, A.G.: Reinforcement Learning: An Introduction. MIT Press (2018)
15. Torabi, F., Warnell, G., Stone, P.: Behavioral cloning from observation. Preprint (2018). arXiv:1805.01954
16. Vaswani, A., Shazeer, N., Parmar, N., Uszkoreit, J., Jones, L., Gomez, A.N., Kaiser, L., Polosukhin, I.: Attention is all you need. Preprint (2017). arXiv:1706.03762
17. Vithayathil Varghese, N., Mahmoud, Q.H.: A survey of multi-task deep reinforcement learning. Electronics **9**(9), 1363 (2020)
18. Watkins, C.J., Dayan, P.: Q-learning. Machine Learning **8**(3-4), 279–292 (1992)
19. Zhang, J., He, T., Sra, S., Jadbabaie, A.: Why gradient clipping accelerates training: A theoretical justification for adaptivity. Preprint (2019). arXiv:1905.11881
20. Zhang, S., Sutton, R.S.: A deeper look at experience replay. Preprint (2017). arXiv:1712.01275

Potential Sales Estimates of a New Store

Jacopo Tozzi and Francesco Guarino

Abstract This work describes a real use cases of a Machine Learning (ML) application in the business context.

The use case concerns the estimation of the potential of new Point of Sale (PoS), where for potential we mean the sales volumes from the second year with respect to the opening date of the contract. For this project, both a *Gradient Boosting* model and a *Convolutional Neural Network*, are used together.

The aim is to support the sales managers engaging new PoS with an automatic tool, which in each year's quarter examines all the Italian active commercial activities and returns the most promising ones.

Keywords Machine learning · Convolutional neural network · Gradient boosting machine

1 Introduction and Problem Definition

To define a Point of Sale (PoS) potential, we introduce first the concept of expected value. The PoS sales expected value is defined as the yearly average sales of a group of similar PoS which have certain common features (walkability, population, position, etc.).

Once we identify the variables conditioning the PoS sales, we can use them to cluster the entire PoS network. Each cluster is then characterized by PoS groups with very similar features. It is reasonable to think the sales variation within the cluster as the inner sales skill of the seller. It turns out that the potential can be defined as the sales volume of a PoS in the case the seller has very good selling skills in the same neighbourhood's condition of other PoS.

J. Tozzi (✉) · F. Guarino
IGT, Rome, Italy
e-mail: jacopo.tozzi@igt.com

A. Masone et al. (eds.), *Optimization and Data Science: Trends and Applications*, AIRO Springer Series 6, https://doi.org/10.1007/978-3-030-86286-2_2

A review of clustering literature is out of the scope of this work, therefore the interested reader is addressed to [1–5] for some of the most recent works on clustering methods and applications.

Formally, for each PoS, the reachable potential is specified as the sales 90th percentile within the same belonging cluster.

We choose the 90th percentile and not the maximum in order to not assign an unrealistic potential to a PoS, which might be due to peculiar and not replicable conditions. In other words, we do not want to assign an outlier as potential sales to a PoS.

The work is structured as follows: Sect. 2 is devoted to present the used databases and features; Sect. 3 details the machine learning (ML) techniques at the basis of the proposed approach; Sect. 4 is related to computational results; finally, Sect. 5 provides the conclusions

2 Geospatial DB Creation and Geospatial Features

As we said in the earlier paragraph, to define similar PoS and identify the variables which most contribute to the sales, we need to describe the area and the demography in the proximity of a PoS. For this reason, we created a geospatial database where we store the coordinates of all Italian activities and points of interest (POI) as underground stops, fuel stations, etc.

2.1 The Geospatial Database

We can essentially group the data into three categories:

- Economic activities: in our commercial activities database we have more than six million of activities on all the Italian soil as for example restaurants, factories, stores, etc. With this information we can describe the economic area in the proximity of a PoS.
- Open street map (OSM) [6]: this data is available in a public database, where users are free to localize any POI. In this way we can consider all the proximity POIs.
- ISTAT data [7]: this database is given by the official Italian Statistical Institute, in which we can find all the demographic features as age, instruction level or average income. The granularity is the census section, which is the lowest census unit. This allows us to explain the demographic area in the proximity of a PoS.

2.2 The Geospatial Features

To describe a new PoS area, we need first its geographic coordinates. Once we localize the PoS, the goal is to understand the peculiarities of the proximity area.

We define two ranges within two radii of 750 m (meter) and 1500 m respectively (Fig. 1). The former is used to depict the nearest features of the PoS, while the second to evaluate the greater area around the immediate PoS proximity. In Table 1 there is a summary of all the variables involved.

For the training we use 20,000 PoS opened in the last years and for each PoS we have associated the features described previously. The activities and POI localized are about 9.7 million. To identify the proximity features we should compute 194 billion distances, so we adopt another strategy.

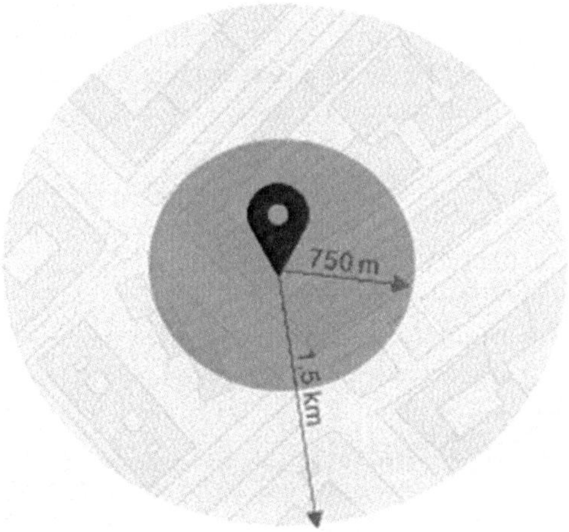

Fig. 1 Example of two radii applied to a potential PoS

Table 1 Table of the explained variables grouped by the three macrocategories

Economic activities	Open Street Map	ISTAT data
Bars and restaurants	Fuel stations	Total population in 1500 m
Commercial activities	Underground and bus stops	Working age population
Entertainment activities	Tourism attractions	Employment rate
Human services activities	Bathhouses	Unemployment rate
Sales of existing PoS	Places of worship	Number of nearest sections
Sales and distance of nearest PoS	Amenities (schools, parks)	Distance of the nearest section
Average income	PoS latitude and longitude	Total city population

We choose the Haversine distance between two points which consider the Earth curvature in the calculation. Since we must compute the distances from commercial activities and POIs within a maximum area of 1500 m of radius centred on the potential PoS, we implement a strategy which allows us to calculate a considerably smaller number of distances.

The idea is to use n centres $C_1, C_2, \ldots, C_n$ (not necessarily PoS), where $n \ll 9.7$ million, on which compute the distances. At this point we must associate at each PoS its nearest centre. We compute only the distances between the PoS and all activities and POI at a distance from the centre less or equal than:

$$D^* = d\left(PoS, C_i\right) + 1500\,\mathrm{m} \tag{1}$$

In this way we calculate all the distances less or equal of 1500 m from the PoS, excluding the majority a priori. The consequent number of distances depends on two factors: the number and position of centres.

For what concerns the number of centres, we need to balance two effects. Specifically, increasing the number of centres, the average distance D^* decreases, consequently there are less distances for activities and POI to compute. On the other side, as the number of centres grows the server memory is closer to its saturation as the number of distances in storage is $dim(C) * Tot_{POI}$, where Tot_{POI} is the total number of activities and POI.

The optimal number of centres found is 8, as one can see in Table 2. In this way we are able to balance the memory usage, the lower number of distances to compute and the computation time. If we add too many centres, the cons, in terms of computation time, are higher than pros.

To choose the best position for each centre we used the clustering algorithm *K-means* on all the 9.7 million activities and POI coordinates. The idea is that the best positioning of the centres is given by the density of the activities and POI. Thus, the average distance D^* becomes smaller.

Table 2 Results of different configurations in centres calculation. The best number of centres is equal to 8, where there is a good balance between memory storage and computation time

	4 centres	8 centres	10 centres	12 centres
Size [GB]	1.8	3.6	4.56	5.38
Computation time [min]	138	95	115	143

3 Proposed Machine Learning (ML) Based Approach

Machine learning techniques are largely used to extract information from large sets and then transforming the information into a comprehensible and compact structure for further use, at the basis of decision support tools dealing with complex decision problems, as shown in [8–11].

The objective is to estimate the PoS sales potential through a model trained over 20,000 PoS already active on the Italian soil. For each PoS we build all proximity variables described previously when the contract has been stipulated with the company. The target variable assigned is the sales volume at the second year from the contract beginning.

To maximize the predictions accuracy, we decide to use, alongside the proximity features, the satellite pictures of the areas around the potential PoS. We take the pictures through the Google "*maps static*" API [12].

Due to the different nature of data for what concerns the satellite pictures, we implement two different models, one stacked with the other, for the two different datasets. The proximity features have been treated as input for a Gradient Boosting Machine (GBM) while the pictures as input of a Convolutional Neural Network (CNN).

3.1 CNN and Satellite Pictures

Specifically, we implement the CNN to create the image embedding in a one-dimension vector, which in turn we concatenate with the proximity features of the Gradient Boosting.

The idea is to exploit the CNN last trained layer as the embedded image vector, where the input is the satellite picture and the target variable the PoS sales. In this way the last layer before the regression neuron will emphasize the most relevant picture features for the sales volumes.

In Fig. 2 we represent the CNN and GBM structures. In Fig. 3 we can see the architecture implemented to store the image requested to Google *maps static* API. We run a batch job in our *Big Data Application* which sends 20,000 requests to the API which in turn responds with the relative pictures stored in our Hadoop infrastructure.

The Convolutional Neural Network is trained using the satellite images as input and the store sales as the target variable.

The aim is to exploit the last dense layer of the trained network to create new features for the meta model. The idea is to construct dense features based on the image patterns and guided by the store sales. The model is built on a pre-trained CCN (InceptionResNetV2) available in *Keras* library [13]. The pre-trained weights are obtained using the *Imagenet* dataset [14].

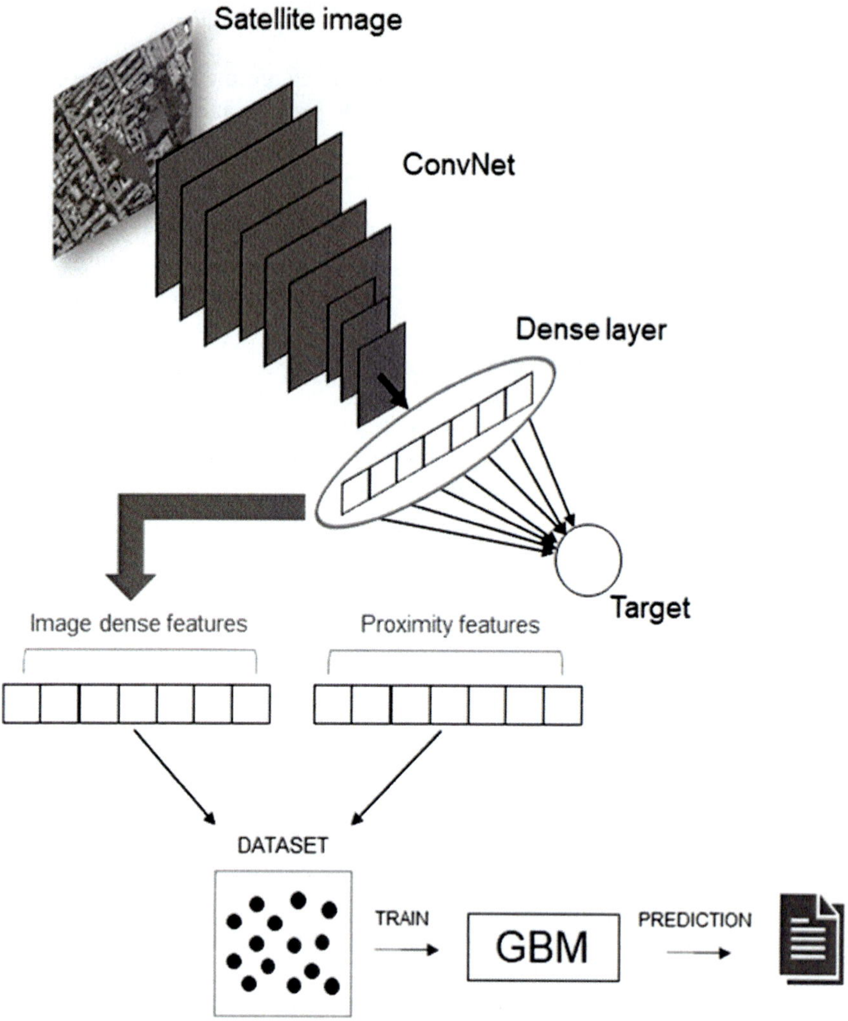

Fig. 2 CNN and GBM models architecture

The pictures size shape is (300, 300, 3), where the first two dimensions indicate the number of pixels while the third one the image channels.

We use the transfer learning technique in four different configurations in the final layers summarised in Table 3. The best result in terms of R^2 (8.5%) has been achieved with one last layer of 128 neurons with *Relu* as the activation function. Moreover, when we implement two final layers, we adopt a middle dropout layer. The dropout rate is equal to 35%. We choose to maximize this metric to understand how much influence the pictures features have on the sales volumes. The R^2 is consistent with the results of CNN applied to the house price challenge [15]. We

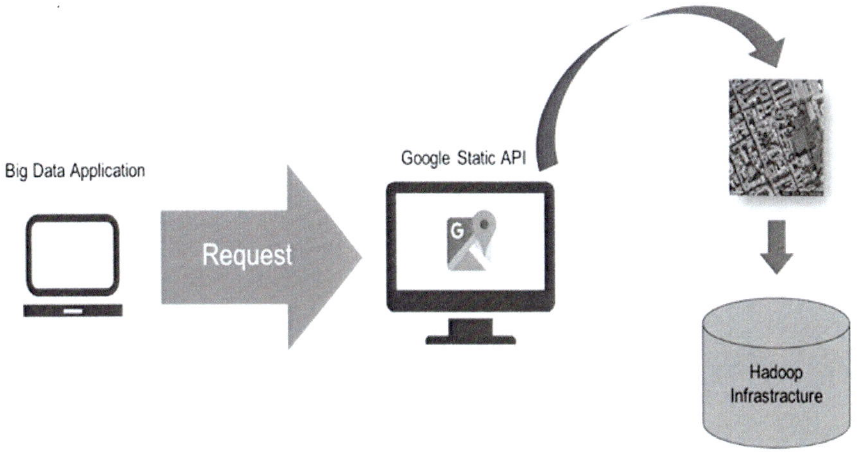

Fig. 3 Google API flow for the satellite pictures collection

Table 3 Results of the different CNN configurations

	1-Layer, 16	2-Layer, 16	1-Layer, 128	2-Layer, 128
R^2	5.3%	3.5%	8.5%	6.2%
Dropout	0%	35%	0%	35%

also apply a decay learning rate (Fig. 4) to train the neural network faster in the first steps and exponentially slower as the number of steps increases.

To train the CNN a mean squared loss function has been minimized using the Adam optimizer.

3.2 GBM for Geospatial Features Based Potential Prediction

Once the CNN is trained, we concatenate the last dense layer to the proximity features. Thus, we can synthesize the image features and use them as input to the gradient boosting model. The target variable is the stores sales, as in the CNN model.

The challenge of this use case is related to the highly unbalanced target distribution (the mean is much greater than the median). To tackle the problem, we minimize the following quantity [16]:

$$E\{\alpha|\boldsymbol{Y} - f(\boldsymbol{X})|I(\boldsymbol{Y} > f(\boldsymbol{X})) + (1-\alpha)|\boldsymbol{Y} - f(\boldsymbol{X})|I(\boldsymbol{Y} \leq f(\boldsymbol{X}))\} \tag{2}$$

$\boldsymbol{Y}$ is the target variable vector, $\boldsymbol{X}$ is the vector of real predictor variables and f is the fitting function. $I(\boldsymbol{Y} \leq f(\boldsymbol{X}))$ and $I(\boldsymbol{Y} > f(\boldsymbol{X}))$ are mutually exclusive indicator variables. The number α is a symmetry factor between 0 and 1, which is used to

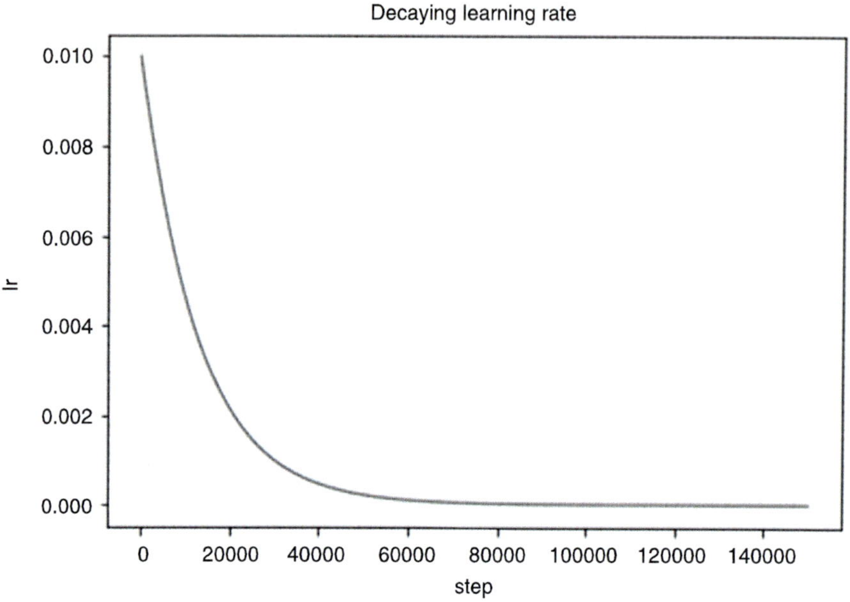

Fig. 4 Plot of the learning rate strategy used in the CNN training

underestimates by α and overestimates by $1 - \alpha$. In general, the function f is the quantile of $\boldsymbol{Y}$.

In our case the best results in terms of R^2 are obtained with $\alpha = 0.5$ which denotes a symmetric underestimating and overestimating in the response. In this case the function f is the median of $\boldsymbol{Y}$ in the tree terminal node.

4 Results of the Proposed ML Approach

After a tenfold cross validation with the *caret* library [17], an R software package [18], we trained a gradient boosting model with 1800 trees, a shrinkage of 0.002, an interaction depth of 5 and 50 as minimum number of observations in the terminal nodes of the trees.

The best R^2 is 52% and we have a MAPE of 0.6. In Fig. 5 we can see the plot of the predicted versus real sales, where we decide to obscure the values on the axis for privacy reasons. For higher values, except for some outliers, the predicted values are more distributed along the bisector.

However, the model is not able to predict values too high, penalising the fewer PoS which perform better. For lower values, we can see that the model basically overestimates the real sales. This is exactly the hard part of this use-case, in which the endogenous variables play a critical role.

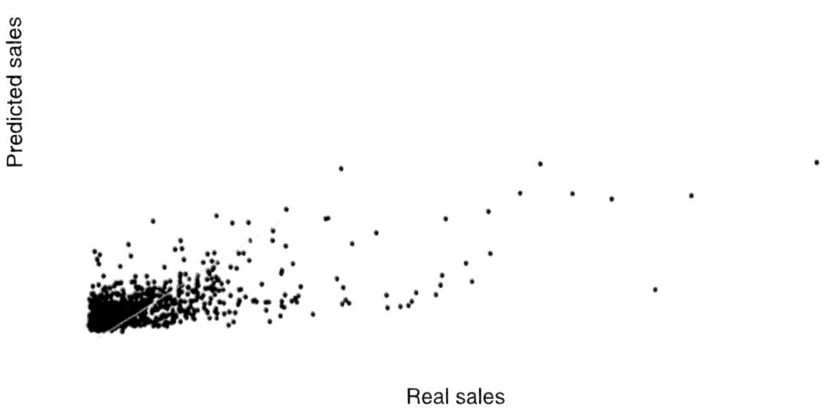

Fig. 5 Plot of predicted versus real sales

5 Conclusions

This use case shows several characteristics that make the analysis challenging. The R^2 obtained indicates that all exogenous variables used explain the 52% of the yearly sales variance. The endogenous variables have still a crucial part in the sales contribution.

A solution might be trying to extract more variables from PoS marketing surveys, which could be good candidates to model the problem endogeneity. We are confident that a good set of surveys variables can improve the R^2 of a certain amount. The endogeneity of the problem would probably continue to take a conspicuous part of the target explanation, but we might reduce its impact in the predictions' accuracy.

The CCN can be improved by training all the weights of a smaller and simpler custom neural network. We used the transfer learning technique because of infrastructure issues, which we count to solve in short times.

References

1. Ahmed, M., Seraj, R., Islam, S.M.S.: The k-means algorithm: a comprehensive survey and performance evaluation. Electronics. **9**, 1295 (2020)
2. Fortunato, S., Hric, D.: Community detection in networks: a user guide. Phys. Rep. **659**, 1–44 (2016)
3. Yang, Y., Wang, H.: Multi-view clustering: a survey. Big Data Min. Anal. **1**(2), 83–107 (2018)

4. Masone, A., Sforza, A., Sterle, C., Vasilyev, I.: A graph clustering based decomposition approach for large scale p-median problems. Int. J. Artif. Intell. **13**, 229–242 (2018)
5. Masone, A., Sterle, C., Vasilyev, I., Ushakov, A.: A three-stage p-median based exact method for the optimal diversity management problem. Networks. **74**, 174–189 (2019)
6. OpenStreetMap contributors. Taken from https://www.openstreetmap.org (2017)
7. ISTAT. Taken from https://www.istat.it/it/archivio/104317 (2011)
8. Olafsson, S., Li, X., Wu, S.: Operations research and data mining. Eur. J. Oper. Res. **187**(3), 1429–1448 (2008)
9. Meisel, S., Mattfeld, D.: Synergies of operations research and data mining. Eur. J. Oper. Res. **206**(1), 1–10 (2010)
10. Boccia, M., Sforza, A., Sterle, C.: Simple pattern minimality problems: integer linear programming formulations and covering-based heuristic solving approaches. INFORMS J. Comput. **32**(4), 1049–1060 (2020)
11. Boccia, M., Masone, A., Sforza, A., Sterle, C.: A partitioning based heuristic for a variant of the simple pattern minimality problem. In: International Conference on Optimization and Decision Science, pp. 93–102. Springer, Cham, September 2017
12. Google Inc. Maps static API. Taken from https://developers.google.com/maps/documentation/maps-static/overview (2021)
13. Chollet, F., et al.: keras. Taken from https://keras.io (2015)
14. Deng, J., Dong, W., Socher, R., Li, L.-J., Li, K., Fei-Fei, L.: Imagenet: a large-scale hierarchical image database. In: 2009 IEEE Conference on Computer Vision and Pattern Recognition, pp. 248–255 (2009)
15. Law, S., Paige, B., Russell, C.: Take a look around: using street view and satellite images to estimate house prices. arXiv. (2019). https://doi.org/10.1145/3342240
16. Kriegler, B., Berk, R.: Small area estimation of the homeless in Los Angeles, an application of cost-sensitive stochastic gradient boosting. Ann. Appl. Stat. **4**, 1234–1255 (2010)
17. Kuhn, M.: Building predictive models in R using the caret package. J. Stat. Software. **28**(5), 1–26 (2008)
18. R Core Team. R: a language and environment for statistical computing. Tratto da. http://www.R-project.org/ (2020)

Sells Optimization Through Product Rotation

Jacopo Tozzi and Francesco Guarino

Abstract This article describes a real use cases of applications of Machine Learning (ML) and Optimization techniques in the business context.

Following the business requests, the Proofs of Concept were developed first and then put into production, rethinking the existing processes and integrating the role of these algorithms in the decision-making processes.

The use case goal is to maximize the overall revenues through the rotation of Vending Machines (VMs) within a network of points of sale (PoS), the VMs sell different products. A new product in a PoS brings a revenue increase in the first weeks. To estimate the expected revenue in the new PoS, in the first n weeks, for all combination of VMs exchange, two ML models have been used.

After the revenue's calculation for all possible exchange, to choose the optimal transfer chain and respect the business constraints, an optimization model has been used. The optimization model has been formulated as a goods transportation problem, where each node (VM position) on the network can be a source or destination node.

Keywords Machine learning · Advanced analytics · Transportation problem

1 Introduction and Problem Description

The company goal is to maximize the vending machines (VMs) sales volume within the network of points of sale (PoS), each VM delivers only one type of product.

The business evidence is that as the number of days that the VM remains in the store increases, (and therefore of the same product), sales decrease.

J. Tozzi (✉) · F. Guarino
IGT, Rome, Italy
e-mail: jacopo.tozzi@igt.com

A. Masone et al. (eds.), *Optimization and Data Science: Trends and Applications*, AIRO Springer Series 6, https://doi.org/10.1007/978-3-030-86286-2_3

A "novelty effect" is assumed within the store, which leads the customer having a greater propensity to purchase the product in the first period of exposure in the PoS.

An initial approach of increasing sales through the rotation of VMs is already carried out by the manager of the physical machine. From the analysis of the data, is possible to see an actual increase in sales volumes in the first n days, the data of all past rotations will be used for training the ML model.

The goal is to create an automated tool that advises managers on optimal rotations, which maximize product sales.

This tool must also provide constraints on movements, for example, from a PoS it cannot be moved more than half of the machines or a maximum of X movements can be carried out.

The first step is to quantify the increase in sales (or the loss in case of still) and analyse the duration of the novelty effect on the PoS.

Assuming not all products suffer the same decline, it has been decided to analyse different curves, each corresponding to a producer (each producer makes products that are very similar to each other). In the data the confirmation of the different effects has been found. In Fig. 1 are three graphs, each belonging to the products of a single producer:

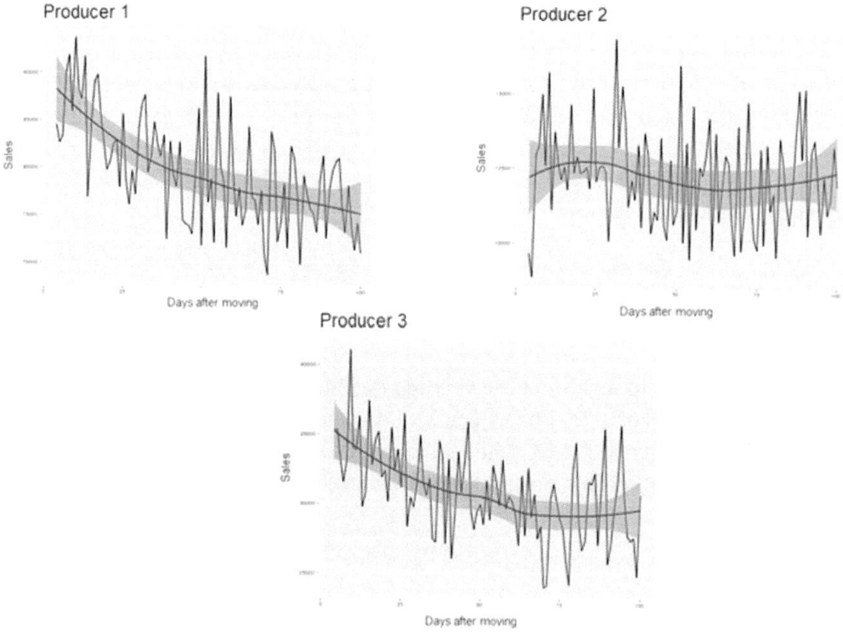

Fig. 1 Sales producer in case of still VMs

For each graph the average product sales are analysed in the following 100 days after the rotation, the averages sales have been calculated on approximately 3500 movements in the last 2 years, to mitigate any seasonal or distorting effects.

As you can see from the graphs, for the different manufacturers, there are different effects. On one hand some producers suffer a more pronounced decline (20%), on the other hand on different producers the time series is almost stationary.

The effect of the decline in overall sales calculated on the first 100 days is about 14%.

Studying the curves in detail, a first slowdown in the decrease in sales occurs after 35 days and in any case most of the decline is concentrated in the first 50 days.

In agreement with the business, the decision was therefore taken to make a VM activated for rotation, after the 35th day of parking in the PoS.

The average effect of the 14% decline can give us an indication of the increase in sales that this model can bring. In fact, in the event of the absence of estimation errors and the "ideal" purchasing behaviour of the customer i.e.: that the customer responds in a positive and constant way to the frequent rotation of the products, a maximum increase in sales of 14% is expected.

Further analyses were carried out to evaluate the effects of rotations on different types of PoS (square meters, position, other products / services inside the PoS).

The differences in the novelty effect of the product in the PoS, are not conditioned only by the type of product / producer, but also by the characteristics of the PoS and its geographical position.

To make a prediction of the sales of a product, moved to a new PoS, a model is needed to consider the interaction between a multitude of variables, which is the reason of using an ML model to make these predictions.

The model has been built to be customized by users according to their requirements or constraints.

First, it provides two series of main constraints, the first is the one that prohibits the movement of VMs, within the same PoS, the second series of constraints, prohibits the movement of a VM containing a specific product on PoS that already has that product, as these shifts would not create the novelty effect.

In addition, there are other more operational constraints, which allow you to configure the total number of VM movements and the maximum number of VMs that can be moved by a single PoS. An additional component is given by adding the calculation of the distances for the movement of VM between PoS in the objective function. In fact, the movement of the physical machine produces logistical costs, so by adding in the objective function a quantity that is a function of the distances between two PoS, this will decrease the gain in relation to the distance, in the case of the choice of the movement.

2 Solution Approach

After the collection of data and information, it has been decided to use two ML models: one for the prediction of sales in case of moving the VM to a new PoS $P^{'}$, (the target variable represents the sum of sales in the first 35 days after moving); the second model will be used to estimate sales in the same 35 days, in the case that the VM already present in $P^{'}$ is not moved.

The difference between these two quantities represents the estimated gain from moving the VM to the new store. In the case of a not profitable shift, the difference between these quantities will assume negative values.

After the calculation of all the possible gains, represented by the deltas of all the possible combinations of displacement of a VM with respect to its stationing, a graph is created where each arc represents the displacement of VM i instead of VM j, the weight associated with arc is equal to the estimated gain in the displacement, therefore a VM will correspond to each vertex. The project is divided into the two phases described below.

The phase 1 is focused on estimating the difference between the sales obtained if the product would be moved in the new PoS (point of sales) in the first n weeks, and the sales which the product would have guaranteed if in the next n weeks, it would not be moved.

Identifying R as the revenue of a vending machine and ΔR as the sales difference in case of VMs exchange, the equation (1) describes the expected revenue in case of moving VMs:

$$\Delta R^{i,j}_{t,t+n} = R^{i,v*}_{t,t+n} - R^{j,v*}_{t,t+n} \tag{1}$$

i = i-th vending machine position, $v*$ = PoS target, n = number of weeks

$R^{i,v*}_{t,t+n}$ estimation (Vending machines rotation)

- A ML model has been built (Random Forest) which estimates the product sales in the new PoS, where the transfer is possible. For the estimation, different information is used, as the product features and the target PoS features (see Sect. 3).
- For each possible product transfer towards a new PoS, the model calculates the sales prediction in the next n weeks.

$R^{j,v*}_{t,t+n}$ estimation (Vending machines still)

- Also, for this indicator a ML model has been implemented, which uses the historical information on the PoS sales, seasonal variables and PoS features (see Sect. 3).
- For each product and PoS, the model calculates the prediction of sales in the case the product is still.

$\Delta R^{i,j}_{t,t+n}$ calculation

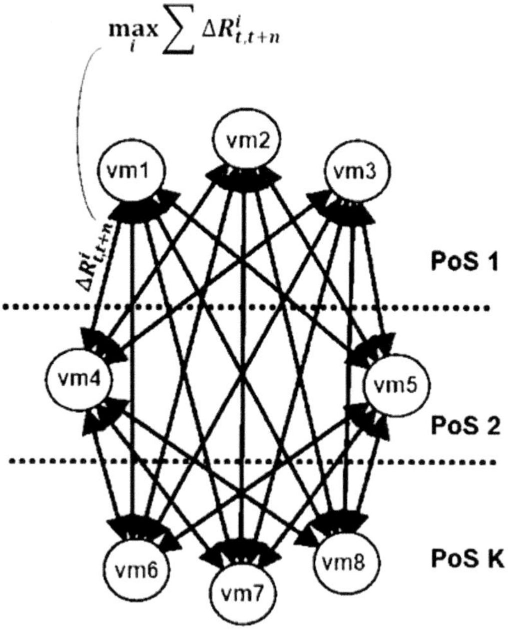

Fig. 2 An example of graph structure, describing a network of K PoS

- Estimation, through the same logics described before, of $R^{i,v*}_{t,t+n}$ and $R^{j,v*}_{t,t+n}$ for the vending machines which are potentially rotated
- All the possible combinations of $\Delta R^{i,j}_{t,t+n}$ are calculated.

In the phase 2, the $\Delta R^{i,v*}_{t,t+n}$ estimated, are used in an optimization model, to select the optimal transfers net on business constraints (each PoS must have always the same number of machines, the same product must not be already present in the target PoS, etc.), the model structure is shown in Fig. 2.

The optimization result is, for each vending machine, the target PoS (which could be also the starting one if the transfer is not convenient).

The model optimizes the vending machines rotations among PoS, basing on two drivers:

- Maximize the sales after a transfer, i.e., the convenience of rotation is assured. The maximization is estimated considering all possible swaps, for which the best solution is feasible together with each available machine
- A series of business constraints has been considered, see constraints from (4) to (11).

3 Sale Prediction of VMs

To predict the sales of still VM on a point of sale, sales of approximately 15,000 VMs in a year has been considered. The target variable represents the sum of sales in the following 35 days after the sampling date. To avoid introducing the novelty effect within the target variable of the stationing, the initial date is randomly sampled between the 50th day starting from the date of movement in the PoS, up to 35 days before the next movement of the VM in another PoS.

In particular, the following variables have been used to estimate the collection of a VM in the next 35 days, in the PoS in which it is already located:

- The sales of the last week
- The sales of the second to last week
- The sales of the third to last and fourth to last week (added together)
- Global median weekly sales
- Standard deviation of the sales
- Number of days since the VM is in the PoS
- Calendar variables (the month in which the estimate is made)
- Day of the year
- Geographic variables (latitude and longitude of the PoS)
- Holidays present in the prediction period

For the sales prediction, after some tuning, the Random Forest model has been chosen.

The model has been trained through cross validation on approximately 12,000 VMs and tested on 3000, below are the results on the predictions.

Comparing the real values with the expected ones, the model has an average estimation error of about 2400 €, predicting about 76% of the variability of the target.

In Fig. 3 there is a comparison between real values and predicted by the model.

The model for the prediction of the sales of a moved VM to a new PoS in the first 35 days after moving, has been trained on about 3000 VMs rotated in the period considered. Only the movements of VMs containing products that are not already present in the arrival points of sale have been evaluated. Compared to the previous model, more information is required to estimate sales. In the previous model the impact of the product in the store in the period before the forecast is known, in the movement model instead it is necessary to estimate the impact of a specific product, and its characteristics compared to the characteristics of the PoS to which it will be moved and of the population that gravity around.

The target variable represents the sum of the sales of the product contained in the VM in the first 35 days within the new PoS.

The forecast variables are the following:

- The variables relating to the new PoS
  - Type of shop

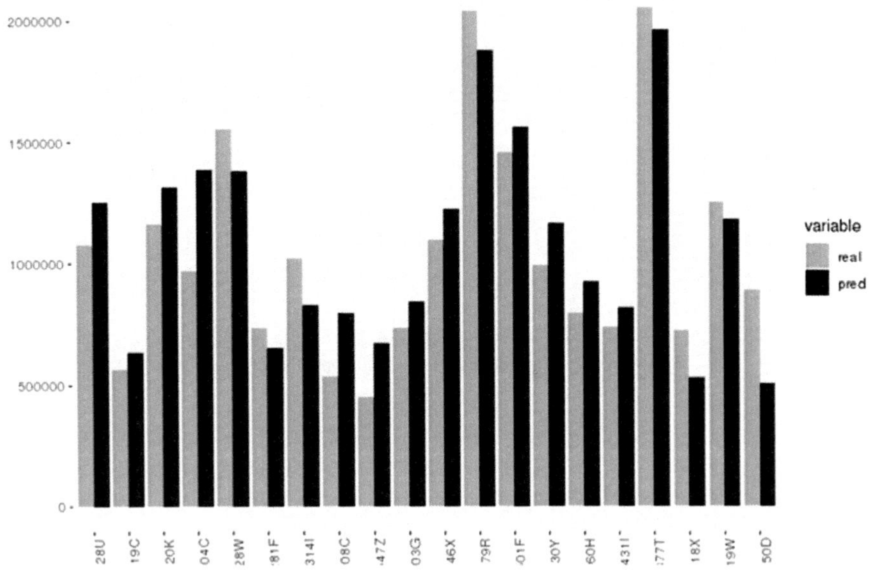

Fig. 3 Example of difference between real and predicted value of VMs sales. The model R-squared is equal to 76%, average error 2400 €

  - Square meters
  - Latitude and longitude of the store
  - Average number of VMs within the PoS
  - Median monthly sales of a VM
  - Sales from other businesses within the PoS
  - The characteristics of the population around the PoS

- The variables relating to the characteristics of the product contained in the VM that will be moved
- Calendar variables (i.e. the month in which the exchange takes place)
- Holidays in the considered period
- Variables relating to the "saleability" of the product, for example the average sales within all the PoS and the positioning with respect to the other products sales.

A Random Forest model has also been used for this prediction.

The model has been tested on approximately 500 VMs moved with the aim of predicting the harvest of the first 35 days.

Comparing the real values with the expected ones (Fig. 4), the model has an average error of about 2600 € going to predict about 64% of the variability of the target variable.

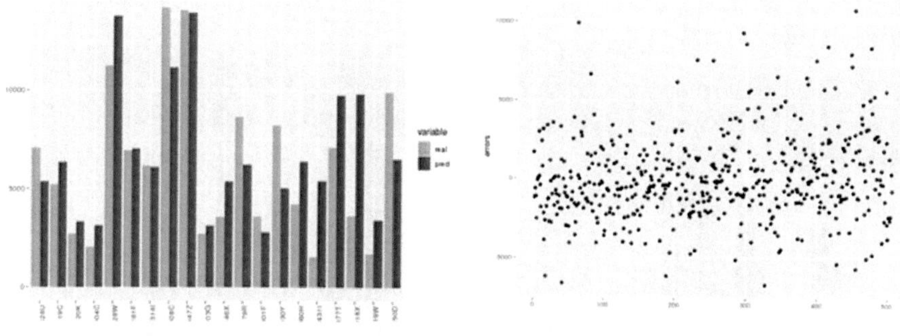

Fig. 4 On the left an example of VMs sales predicted and real sales values in the case of moved VMs, on the right the residuals of estimation errors. The R-squared is equal to 64% and the average prediction error is about 2600 €

4 Problem Formulation

After estimating the sales predictions in the case of moving or stationary VMs, the goal is to make the optimal movements of the VMs within the PoS network, meaning the ones that maximize the sales in the following 35 days.

The problem is thought as an optimization model on directed graph and adapted for the modelling of the movement of goods within a network of PoS.

In the graph $G(V, E)$ the set of vertices V represents the VMs within the network of PoS, therefore $|V|$ will be equal to the number of VMs considered.

The generic arc e which belongs to the set E, represents the displacement of the VM u in v with $u, v \in V$.

Since the displacement of the VM u in the position of the VM v is not symmetrical with respect to the exchange of u and v, a directed graph is considered.

The weight to be attributed to each arc $a_{i,j}$, which corresponds to the displacement of VM i in j, will be equal to the following formula:

$$\Delta R^{i,j}_{t,t+n} = R^{i}_{t,t+n} - R^{j}_{t,t+n} \quad (2)$$

where R in (2) represents the estimated sales of the VM in the PoS, i = i-th vending machine position, j = j-th vending machine position, n = number of weeks.

Since a store can contain several VMs inside it, and only VM movements to other PoS are foreseen, a Z partition of the vertices $Z_1, \ldots, Z_n$ is therefore performed, where the generic partition Z_i includes all the vertices (VM) within a PoS, with $Z_i \cap Z_j = 0 \; \forall \; i, j$. The cardinality of Z will therefore be equal to the number of PoS in the network.

A further partition of the set V is carried out according to the type of product within the VM. A partition S of the vertices $S_1, \ldots, S_m$ has been performed, where the generic partition S_k includes all vertices (VM) that sell the same product k, with

$S_h \cap S_k = 0 \; \forall \; k, h$. The cardinality of S will be equal to the number of different products within the VMs.

The binary variable x_{ij} is a decision variable, which takes the value 1 if VM i is moved to the position of VM j, 0 otherwise.

Considering N VMs, H products and K PoS, the optimization model is shown below:

$$\text{maximize} \sum\nolimits_{i,j=1}^{N} R_{ij} * x_{ij} \tag{3}$$

$$\sum\nolimits_{j=1}^{N} x_{ij} \leq 1 \qquad \forall i = 1, \ldots, N \tag{4}$$

$$\sum\nolimits_{i=1}^{N} x_{ij} = \sum\nolimits_{i=1}^{N} x_{ji} \qquad \forall j = 1, \ldots, N \tag{5}$$

$$\sum\nolimits_{i,j=1}^{N} x_{ij} \leq B \qquad \forall i, j = 1, \ldots, N \tag{6}$$

$$\sum\nolimits_{\substack{i \in S_h \\ j \in S_h}} x_{ij} = 0 \qquad \forall h = 1, \ldots, H \tag{7}$$

$$\sum\nolimits_{\substack{i \in Z_k \\ j \in Z_k}} x_{ij} = 0 \qquad \forall k = 1, \ldots, K \tag{8}$$

$$\sum\nolimits_{\substack{i \in Z_k \\ j \notin Z_k}} x_{ij} \leq \left\lfloor \frac{|Z_k|}{2} \right\rfloor \qquad \forall k = 1, \ldots, K \tag{9}$$

$$\sum\nolimits_{\substack{i \in S_h \\ j \in Z_k}} x_{ij} \leq 1 \qquad \begin{matrix} \forall k = 1, \ldots, K \\ \forall h = 1, \ldots, H \end{matrix} \tag{10}$$

$$x_{ij} \begin{cases} 1 \; if \; the \; vm \; i \; is \; trasnferred \; in \; j \\ 0 \; otherwise \end{cases} \tag{11}$$

S *vm partition with the same products*
Z *vm partition in the same PoS*

The R_{ij} in the objective function correspond to the $\Delta R_{t,t+n}^{i,j}$ previously calculated as the difference of the two models' outputs.

If the product of VM i is already in the PoS of VM j, by convention $R_{ij} = 0$, as there is no novelty effect within the PoS. Therefore, it is potentially possible to move a product within the PoS with the same product already present, but locally it is not convenient, it can only be in the case of a configuration that is able to maximize the overall solution.

The binary variable X_{ij} is a decision variable, which takes the value 1 if VM i is moved to the position of VM j, 0 otherwise.

Model constraints are the followings: (4) a maximum of 1 VM can be moved from its position, (5) the flow conservation constraint, guarantees that the solution maintains the initial configuration of the number of machines per PoS, using this constraint, the only possible operations for the movement are the exchange between two VMs, or a circular exchange chain, (6) this constraint assures us that a maximum of B exchanges are possible on the PoS network, (7) VMs containing the same product cannot be swapped, (8) exchanges in the same PoS are not allowed, (9) a maximum of half of the VM present in a PoS can be moved (configurable parameter), (10) more than one product of the same type cannot be moved to a new PoS.

To consider the distances between the movements of VM between PoS and penalize the movements between two far PoS, it is possible to modify the objective function (3) in the following way:

$$\text{maximize} \sum\nolimits_{i,j=1}^{N} R_{ij} * x_{ij} - \alpha * \left(D_{ij} * x_{ij}\right) \tag{12}$$

D_{ij} are the distances between two VMs, by setting the parameter α it is possible to increase or decrease the penalty on the moving distances.

In the following results (12) is used as maximization function.

A review of routing problems is out of the scope of this work, therefore the interested reader is addressed to [1–4] for some of the most recent surveys and applications.

5 Computational Results

The data used for training the ML models and the optimization model are collected in HDFS [5], the distributed file system in the Big data architecture.

A first processing is carried out through Spark, running the ETL (extract, transform, load) processes on millions of records, to process the aggregates that will be input to the models listed above.

Machine learning models (Random Forest) have been trained using R software [6], the package *caret* [7] has been used. The optimization model has also been implemented through R, for the construction of the constraints and the objective function the package *ompr* [8] has been used, as the solver instead the package linux *Symphony* [9], which can be directly called from R.

The model is run quarterly and choose the optimal movement chains, which maximize the total revenue on the PoS network, in Fig. 5 there is an example of the optimal exchange chains suggested by the model.

Table 1 is an example of a model run, on a sub-sample of PoS and VM.

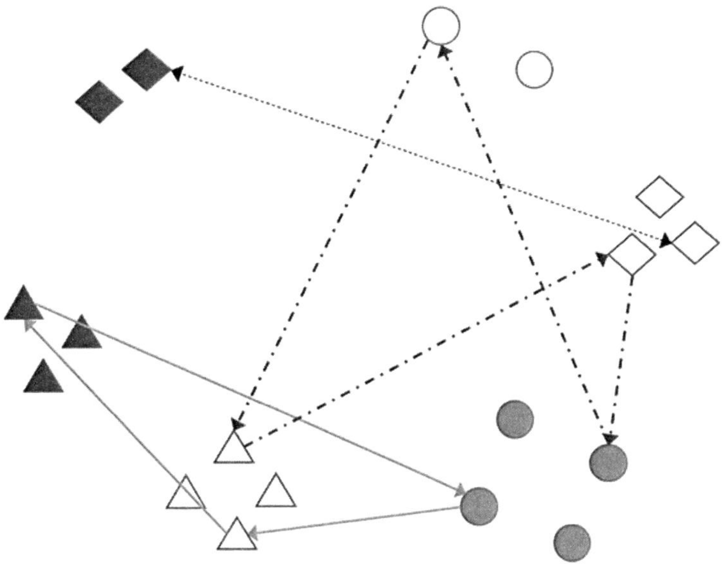

Fig. 5 Model's run example on 6 PoS (different shapes) and 18 VMs

Table 1 Input parameters and output results of a run example

Number of PoS	50
Number of VM	199
Number of different products	53
Alpha (α)	30
Sales increase to 35 days	267,183
Number of VM moved	91

6 Conclusions

Using this model, it is possible to optimize the handling of different products, increasing sales on the PoS network.

It is expected that the estimated increase in sales due to the application of the model, including estimation errors and a lower customer response to a frequent novelty effect, is lesser then 14%, which corresponds to the novelty effect that has been studied in the preliminary analysis on real data.

For a better estimate of the optimal movements, it will be possible to improve the performance of the ML models, through the retraining of the models after the suggested movements and enriching the input data with further variables.

Subsequent developments of the optimization model are possible, specifically, the limitation of the exchange chains length, to make the model more logistically efficient.

To limit the exchange chain length, it is necessary to enumerate the movements in the chain, this is possible by introducing the Subtour Elimination Constraints (STE), which however reduce the efficiency of the model by increasing its complexity.

Other developments are possible, adding further constraints, which allow the VMs manager to obtain flexible outputs with respect to the desired one.

References

1. Archetti, C., Speranza, M.G.: A survey on matheuristics for routing problems. EURO J. Comput. Optim. **2**, 223–246 (2014)
2. Boccia, M., Masone, A., Sforza, A., Sterle, C.: A column-and-row generation approach for the flying sidekick travelling salesman problem. Transp. Res. Part C. **124**, 102913 (2021)
3. Castillo-Salazar, J., Landa-Silva, D., Qu, R.: Workforce scheduling and routing problems: literature survey and computational study. Ann. Oper. Res. 1–29 (2014)
4. Sinha Roy, D., Masone, A., Golden, B., Wasil, E.: Modeling and solving the intersection inspection rural postman problem. INFORMS J. Comput. (2021). https://doi.org/10.1287/ijoc.2020.1013
5. Apache Software Foundation. Apache Hadoop. Taken from https://hadoop.apache.org
6. R Core Team. R: A language and environment for statistical computing. Taken from http://www.R-project.org/ (2020)
7. Kuhn, M.: Building predictive models in R using the caret package. J. Stat. Software. **28**(5), 1–26 (2008)
8. Schumacher, D. OMPR: R package to model mixed integer linear programs. Taken from https://github.com/dirkschumacher/ompr (2020)
9. Symphony Contributors. Symphony MIP Solver. Taken from https://projects.coin-or.org/SYMPHONY (2019)

Part II
Healthcare

Gathering Avoiding Centralized Pedestrian Advice Framework: An Application for Covid-19 Outbreak Restrictions

Veronica Dal Sasso and Valentina Morandi

Abstract Due to the COVID-19 pandemic, the focus on everydays mobility has been shifted from traditional means of transport to how to safely commute for work and/or move around the neighbourhood. Maintaining the safe distance among pedestrians becomes crucial in big pedestrian networks. Looking at personal goals, such as walking through the shortest path, could lead to congestion phenomena on both roads and crossroads, violating the imposed regulations. We suggest a centralized multi-objective approach able to assign alternative fair paths for users while maintaining the congestion level as low as possible. Computational results show that, even considering paths that are not longer than 1% with respect to the shortest path for each pedestrian, the congestion phenomena are reduced of more than 50%.

Keywords Pedestrian route assignment · Gathering avoidance · Covid social routing

1 Introduction

The steady progress towards a globalized world has, in the last decades, reduced the impact of distances. While, in the past, people were born, grew and spent all their lives in the same city, travelling has recently become more popular. The need to commute to reach the workplace consistently increased. As a consequence, the use of public and private means of transportation also increased, before the sudden drop due to the COVID-19 pandemic at the beginning of 2020. Before that, it was common to spend a lot of time travelling, ending up in traffic jams as every car

V. Dal Sasso (✉)
Optrail, Rome, Italy
e-mail: veronica.dalsasso@optrail.com

V. Morandi
Free University of Bozen—Bolzano, Faculty of Science and Technology, Bolzano, Italy
e-mail: Valentina.Morandi@unibz.it

A. Masone et al. (eds.), *Optimization and Data Science: Trends and Applications*, AIRO Springer Series 6, https://doi.org/10.1007/978-3-030-86286-2_4

driver tried to achieve his/her personal optimum, or being packed in overcrowded buses or trains. Now all this appears freezed, while the focus on everyday mobility is shifted on how to safely move around the neighborhood by foot. Walking also replaced public transports whenever possible, as people feel safer in the open air than in buses or metros. This means, however, that the choice of the path to walk through is seen in a new light, because keeping social distance is crucial also in the open air to reduce the risk of COVID-19 infections. Clearly, users tend to look at personal goals, such as minimizing the length of their path, while being oblivious of others' presence. This behaviour usually is self-defeating at keeping low rates of pedestrian congestion on the streets. On the other hand, decisions could be deferred to a centralized entity, which has a global view of the demand and capacity on the streets. Thus, all users could gain something in terms of experienced congestion phenomena.

The presented methodology aims at ensuring social distancing, and hence avoid congestion, among people walking on the city centres streets. When the walking activity is motivated by the need of reaching a destination, most people would choose the shortest path available. Thus, quite intuitively, the users' satisfaction drops fast with the increase in the length of paths. However, a small, nearly unnoticeable, increase in walking time could avoid congestion using a centralized approach.

Thus, the aim of this paper is to show how paths may be assigned to pedestrians, balancing the opposing goals of minimizing the increase on the length for users and minimizing the capacity excess on both the streets and crossroads/squares. We recall that a path is considered eligible if it is no longer than a percentage of the shortest path, guaranteeing fairness among users.

The problem of route assignment for pedestrians fits in the frameworks thoroughly described in the literature review Sect. 1.1. In Sect. 2, the gathering avoiding pedestrian routing model is presented. In Sect. 3, the result of thorough experiments assessing the performance of the model on real road networks is shown. Finally, in Sect. 4, some concluding remarks and future developments are provided.

1.1 Literature Review

The centralized traffic assignment idea was already mentioned by [18], where, speaking about vehicular traffic, the author makes the distinction between user equilibrium and system optimum. User equilibrium is a traffic assignment in which each user decides on its own the best route to follow. Conversely, the system optimum is the traffic assignment in which the total travel time is minimized, without considerations on the behaviour and fairness among users. The deterioration of the overall solution in implementing the user equilibrium versus the system optimum is known in literature as the price of anarchy (see [12] and [14] for details and mathematical background). A compromise solution between these two principles can offer interesting insights, leading to a win-win situation for all people involved. In fact, a myopic view of the problem from the perspective of the single user can lead to choices which, once put into practise, prove to be sub-optimal. As an example, let's consider drivers commuting every day to and from the workplace. Each of them is prone to choose the shortest path, in terms of distance, travelling times or both. But, if their paths intersect, the level of congestion on the streets will increase the travelling times, reducing the effectiveness of the user's choice. Even using real-time information on traffic, users' decisions simply result in a shift of congestion from previously congested roads to other roads. On the contrary, by choosing a different and at first sight less favourable path, the gain for the user may be substantial. As highlighted in [6], however, the majority of users is not willing to act socially but instead selfishly, especially if the cost of acting socially is high [9]. Different approaches for balancing user equilibrium and system optimum have been investigated (see [17] and [13] for comprehensive reviews), spanning different ways of transportation. The bounded rational user equilibrium differs from the user equilibrium in the fact that the users are not completely free to choose their best route. In fact, a number of paths for each user can be considered according to the so-called indifference travel time band (i.e. a collection of paths such that, picked one, the users would not feel the desire to change it). Details on the bounded rational user equilibrium can be found in [11] and in [19]. On the other hand, the constrained system optimum minimizes the total travel time trying to limit the unfairness among users by limiting the set of eligible paths only to those that are not longer than a small percentage with respect to the best choice. This centralized approach inspired this work. The first attempt to formulate the constrained system optimum, as a convex non-linear problem, can be found in [10]. For theoretical bounds on the price of anarchy we refer to [15]. Given the difficulty of handling non-linear latency functions, a first attempt to use a linear programming model to solve the constrained system optimum traffic assignment problem is proposed in [1] and later in [4]. Since the number of eligible paths is exponential, in [2] a fast and reliable heuristic algorithm to solve big road networks is proposed. In all presented approaches the set of eligible paths is generated a priori. Once the flow is routed on paths, it could happen that the experienced unfairness is much higher than the one evaluated a priori. To overcome the issue, in [5] two constrained system optimum formulations directly controlling the real experienced unfairness are presented. None of the presented approaches take into account the arc congestion level as a penalty for the system objective. The first attempt to embed arc congestion reduction techniques has been proposed in [3], where a constrained system optimum with the aim of lowering congestion on worst congested arcs is proposed.

A different approach which can be pursued to combine user equilibrium and system optimum is to formulate the problem as a multi-objective model. An example in the context of Air Traffic Management can be found in [8], where the authors assess the viability of incorporating single airlines preferences within a collaborative European framework. The main goal of the model presented is to ensure that air space capacity over Europe is never exceeded, while trying to accommodate stakeholders' requirements. Indeed, airlines request to the Air Traffic Manager trajectories and departure times for their flights, but usually they are not willing to share which policies lead to such choices. Hence, the requests may be disguised and may vary from one airline to the other, on the basis of the airlines' target. In order to adapt each airline's demand to the global objective of reducing costs, while ensuring the congestion level of air space sectors is below a set capacity, the model tries to minimize the deviation from the preferred trajectories, both in terms of delays and routing, and to minimize the total costs of traversing air sectors. The outcome of the model is a set of Pareto optimal solutions, among which one solution may be selected by further considerations on the fairness between the different stakeholders.

2 The Gathering Avoiding Pedestrian Routing Model

In this Section, we present the Gathering Avoiding Centralized Pedestrian Routing problem (GA-CPR). The model assigns paths to users in order to balance two opposing objectives, i.e. minimizing the increase in path length and minimizing the excess in the capacity. Hence, the problem is formulated as a multi-objective model. The fairness between users is embedded into the model by selecting as eligible paths only those that are not too unfair for users.

In the following, we will assume that the flow of pedestrians is constant. This allows us to neglect the time dimension of the problem and consider a static demand on the network. As assessed by [16], this modelling choice holds when rush hour time slots are considered.

Let $G = (V, A)$ be a directed network, where V represents the set of vertices and $A \subseteq V \times V$ the set of arcs. The set of arcs represents the set of roads in the pedestrian network while the set of vertices represents the set of crossroads between roads in set A. Each vertex $h \in V$ is associated to a capacity cap_h representing the number of pedestrians that can transit through the vertex without causing gathering phenomena, and a traversing time t_h. Similarly, each road is associated with a capacity cap_{ij}, representing the number of pedestrians that can walk through that road segment without gatherings, and a traversing time t_{ij}. The requests for walking through the pedestrian area are collected and the demand of all pedestrians going from the same origin to the same destination are consolidated in an Origin-Destination (briefly OD). The set C represents the set of OD pairs, each associated to an origin $O_c \in V$, a destination $D_c \in V$, and a pedestrian demand rate d_c from O_c to D_c. t_{cSP} indicates the traversing time while walking along the shortest path between O_c and D_c. Only paths that are similar to the shortest path in terms of length are given as input to the model for each OD pair, as it is necessary that the pedestrians feel that the proposed paths are among the best available. In details, we consider as eligible only paths whose relative excess in walking time with respect to the shortest path are within the fairness percentage ϕ. The set of eligible pedestrian paths from origin to destination for each OD pair $c \in C$ is denoted by K_c^{ϕ}. For each $k \in K_c^{\phi}$, let t_{ck} be the time needed to go across path k. The indicator parameter a_{ij}^{kc}

has value 1 if path $k \in K_c^\phi$ traverses arc $(i, j) \in A$, while it has value 0 otherwise. Details on the generating paths algorithm are discussed and provided in [1].

Variables x_{ij} represent the total pedestrian flow on arc $(i, j) \in A$, while variables σ_{ij} represent the excess of flow with respect to arc capacity on arc (i, j). Similarly, variables δ_h indicate the excess of flow traversing a certain vertex $h \in V$ and variables z_h indicate the excess of flow traversing a certain arc $(i, j) \in A$. Moreover, a number of variables are related to each path. Variables y_{ck} represent the pedestrian flow of OD pair $c \in C$ routed on path $k \in K_c$.

The objective functions of the GA-CPR model are denoted by

$$\tau(\phi) = \sum_{c \in C} \sum_{k \in K_c^\phi} \frac{t_{ck}}{t_{cSP}} y_{ck} \tag{1}$$

$$\eta(\phi) = \sum_{(i,j) \in A} \frac{t_{ij}}{cap_{ij}} \sigma_{ij} + \sum_{h \in V} \frac{t_h}{cap_h} \delta_h \tag{2}$$

Objective (1) records the total relative increase in walking time of pedestrians on paths with respect to the shortest one. It depends on the walking time unfairness parameter ϕ. Objective (2), also depending on ϕ, records the total relative excess of capacity for arcs and nodes weighted by the traversing time.

The GA-CPR bi-objective model follows:

$$\min \quad \tau(\phi), \eta(\phi) \tag{3}$$

$$d_c = \sum_{k \in K_c^\phi} y_{ck} \qquad c \in C \tag{4}$$

$$x_{ij} = \sum_{c \in C} \sum_{k \in K_c^\phi} a_{ij}^{ck} y_{ck} \qquad (i, j) \in A \tag{5}$$

$$z_h = \sum_{(i,j) \in A | j=h} x_{ij} \qquad h \in V \tag{6}$$

$$\sigma_{ij} \geq x_{ij} - cap_{ij} \qquad (i, j) \in A \tag{7}$$

$$\delta_h \geq z_h - cap_h \qquad h \in V \tag{8}$$

$$x_{ij} \geq 0 \qquad (i, j) \in A \tag{9}$$

$$z_h \geq 0 \qquad h \in V \tag{10}$$

$$\sigma_{ij} \geq 0 \qquad (i, j) \in A \tag{11}$$

$$\delta_h \geq 0 \qquad h \in V \tag{12}$$

$$y_{ck} \geq 0 \qquad c \in C, k \in K_c. \tag{13}$$

Constraints (4) ensure that the pedestrian demand d_c of OD pair $c \in C$ is routed on paths in K_c^{ϕ}. Constraints (5) set the flow on an arc equal to the sum of the flows on the pedestrian paths traversing the arc. Constraints (6) set the inflow of a vertex to be equal to the sum, among arcs entering in the vertex, of their pedestrian flows. Constraints (7) set σ_{ij}, for each arc $(i, j) \in A$, greater or equal to the excess of flow x_{ij} with respect to the arc capacity cap_{ij}. Notice that, because of constraints (11) and objective function (3), variable σ_{ij} assumes value 0 whenever the arc capacity is not exceeded and $x_{ij} - cap_{ij}$, otherwise. Similarly, constraints (8) set δ_h greater or equal to the excess of flow z_h in vertex $h \in V$ with respect to the vertex capacity cap_z. Together with constraints (12) and objective function (3), variable δ_h assumes value 0 if the vertex capacity is not exceeded and $z_h - cap_h$, otherwise. Finally, constraints (9)–(13) define the domain of the decision variables.

2.1 Solution Method

A number of methodologies can be used to tackle multi-objective optimization. We will refer to [7] for a comprehensive review of such methodologies. We decided to use the linear combination of weights as solution method since, as assessed by [7], the main advantages of this method are its simplicity and its efficiency (from a computational point of view). However, its main drawback is the determination of the appropriate weight coefficients to be used in the final objective function. In fact, the choice of weights is crucial in determining the solution. Thus, the choice has to be made by the decision maker carefully considering the real-world problem characteristics.

The two objective functions $\tau(\phi)$ and $\eta(\phi)$ introduced in the previous paragraph are weighted according to the importance parameters α and 1-α, respectively, with $\alpha \in [0, 1]$. The resulting objective function is:

$$\min \quad \alpha \sum_{c\in C} \sum_{k\in K_c^{\phi}} \frac{t_{ck}}{t_{cSP}} y_{ck} + (1-\alpha)[\sum_{(i,j)\in A} \frac{t_{ij}}{cap_{ij}} \sigma_{ij} + \sum_{h\in V} \frac{t_h}{cap_h} \delta_h]$$

.

In Sect. 3, we will also analyze the impact of the choice of α parameter on the optimal solution. The aim is to give an overview of the impact of the two objective functions on different policy maker choices.

3 Computational Results

A benchmark of 4 map-based instances has been used in a computational study to assess the performance of the presented model. Instances are generated by randomly draw coordinates for origins and destinations of pedestrian flows from four Italian cities, namely Brescia, Bolzano, Rome and Vicenza. Arc walking times are obtained using Graphhopper and arc capacities are obtained by dividing the real distance on map by the safety walking distance of 2m imposed by Covid-19 regulations. Vertices' capacity is obtained as a percentage of the entering arcs capacity and vertices' traversing time is obtained randomly within a short time windows. OD pairs' demands are generated as a percentage of the capacity of the arcs exiting the origin. The 4 tested networks have 50 nodes with approximately 2500 arcs and 25 OD pairs, each one with a different demand. All instances can be found at: https://valentinamorandi.it/research-outcomes/. For each instance, a traffic assignment has been found using a restricted path set with ϕ values in $\{0.01, 0.05, 0.1, 0.15, 0.2\}$ and α values in $\{1, 0.9, 0.7, 0.5, 0.3, 0.1, 0\}$. For each instance, we compute also the user equilibrium (in which each passenger goes on their shortest route) as a matter of comparison. It is obtained by setting $\alpha = 1$ and $\phi = 0$. In total, we obtain 36 traffic assignments for each instance. The model is solved using CPLEX 12.6.0 on a Windows 64-bit computer with Intel Xeon processor E5-1650, 3.50 GHz, and 64 GB RAM. Results for the GA-CPR model are presented and discussed in Sect. 3.1.

In the following all the computed and collected statistics are defined.

- **Congestion distribution**
  - $\bar{\sigma}$: average relative excess of flow with respect to the arc capacity, i.e. $\frac{1}{|A|} \sum_{(i,j)\in A} \frac{\sigma_{ij}}{cap_{ij}}$.
  - $\bar{\delta}$: average relative excess of flow with respect to the node capacity, i.e. $\frac{1}{|V|} \sum_{h\in V} \frac{\delta_h}{cap_h}$.
  - $\lambda_{=0}$: Percentage of arcs and nodes with $\frac{\sigma_{ij}}{cap_{ij}} = 0$ or $\frac{\delta_h}{cap_h} = 0$ w.r.t. the total number of arcs and nodes.
  - $\lambda_{0<\ldots<0.25}$: Percentage of arcs and nodes with $0 < \frac{\sigma_{ij}}{cap_{ij}} < 0.25$ or $0 < \frac{\delta_h}{cap_h} < 0.25$ w.r.t. the total number of arcs and nodes.
  - $\lambda_{\geq 0.25}$: Percentage of arcs and nodes with $\frac{\sigma_{ij}}{cap_{ij}} \geq 0.25$ or $\frac{\delta_h}{cap_h} \geq 0.25$ w.r.t. the total number of arcs and nodes.
- **User experience for each OD pair** $c \in C$
  - $U^{ck} = \frac{t_{ck}-t_{SP}}{t_{SP}}$: user walking unfairness with respect to fastest path for the OD pair.
  - $\bar{U} = \frac{1}{\sum_{c\in C} d_c} \sum_{c\in C} \sum_{k\in K_c^{\phi}} y_{ck} U^{ck}$: weighted user walking unfairness.
- **Network statistics**
  - T: percentage increase of total walking time with respect to user equilibrium (UE).
  - Σ: reduction of the total walking time with exceeded capacity on arcs with respect to UE.
  - Δ: reduction of the total walking time with exceeded capacity on nodes with respect to UE.

3.1 Performance of the Model

In this section, we summarize the results obtained. Multi-objective models are interesting as they are able to show how contrasting goals are affected by each other. By increasing parameters ϕ and α we are, respectively, assessing the impact of loosening the constraint on users' fairness and of focusing on the reduction of congestion.

Quite intuitively, the walking time increases both when more importance is given to the reduction of congestion and when longer routes are made available (see Fig. 1). It is interesting to notice how, by just allowing a small freedom in the increase of the paths length, there is a considerable drop in the total time spent walking on congested nodes or arcs. Figure 2 represents exactly this: the chart on the left shows how, by just adding those paths that are a 1% longer than the shortest path, the reduction of walking time across congested nodes drops by 50%. There is not much improvement, however, when longer paths are allowed. Another interesting results is that there is not much difference when the value of α changes (except, of course, when $\alpha = 1$: when no weight is given to the congestion, there is no incentive in diverting from the shortest path). On the right, a similar chart shows an analogous result for the walking time on congested arcs. Notice that, in this case, the drop is more substantial when longer paths are allowed.

Figure 3 shows how the average user unfairness, calculated as shown in the previous section, increases when longer paths are allowed. It clearly depicts the advantage of considering two contrasting objectives, as by varying the value of α the unfairness can be kept under control.

Figure 4 remarks what is already shown in Fig. 1 but from another point of view. Instead of the total walking time on congested arcs and nodes, we are looking at the excess of flow. Congestion on nodes and arcs drops considerably when paths just 1% longer than the shortest paths are taken into account. There is not much

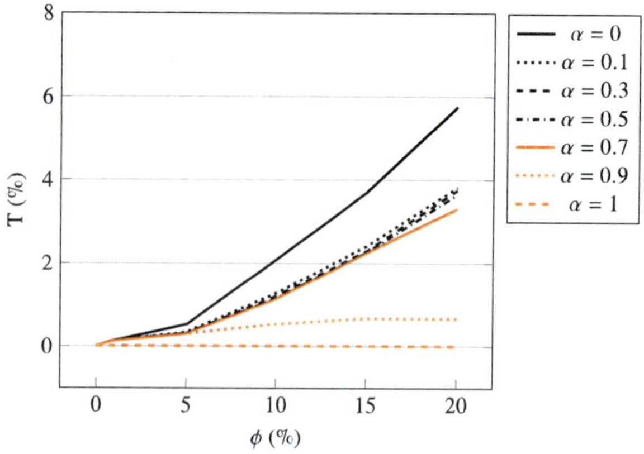

Fig. 1 Increase in total walking time T, as a function of ϕ

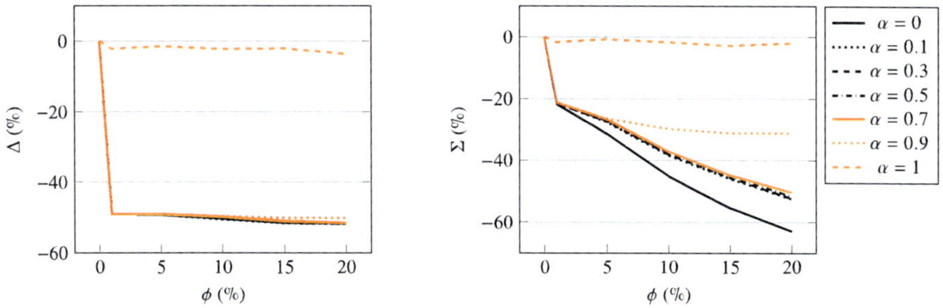

Fig. 2 Decrease of walking time with exceeded capacity on nodes Δ (on the left) and on arcs Σ (on the right), as a function of ϕ

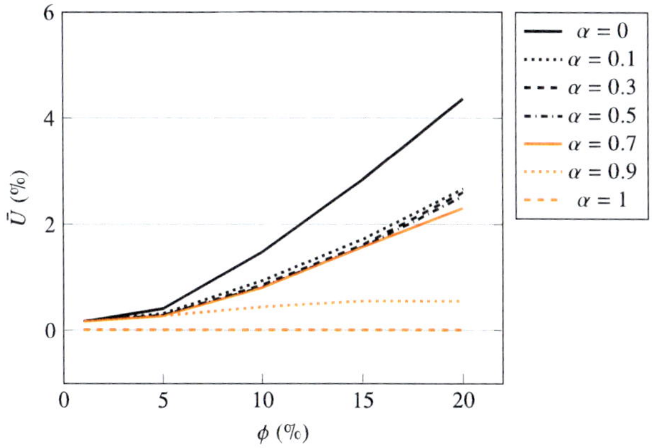

Fig. 3 Average $\bar{U}$ user walking unfairness, as a function of ϕ

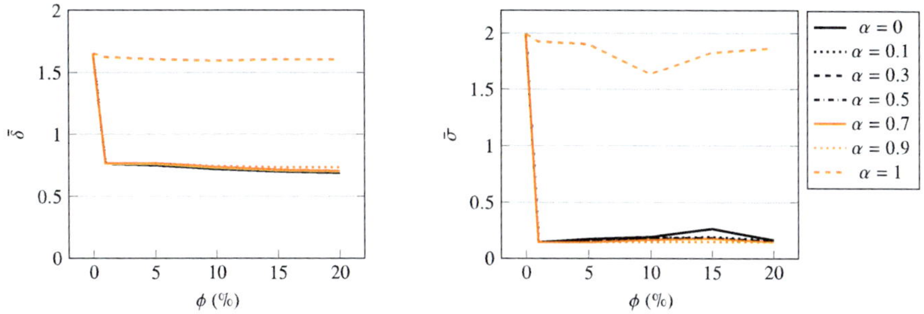

Fig. 4 Average node congestion $\bar{\delta}$ (on the left) and arc congestion $\bar{\sigma}$ (on the right), as a function of ϕ

improvement, instead, in choosing longer paths, as expected. Figure 5 shows the congestion on arcs and nodes at the extreme points of the dominant solutions, i.e. the solutions of the multi-objective model obtained by setting α either 1 or 0. The percentage of non-congested arcs and nodes (the striped region) when only the total

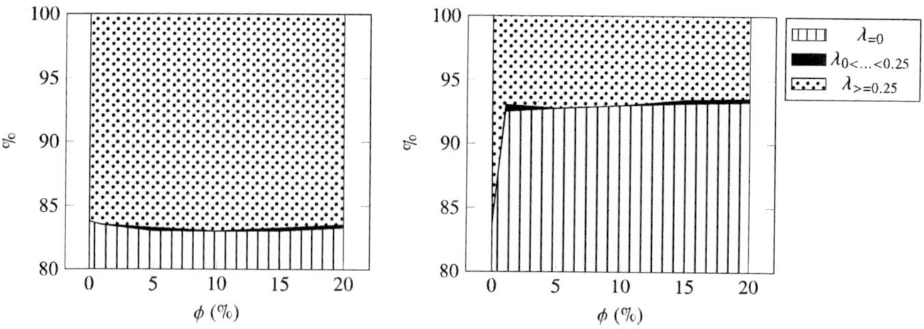

Fig. 5 Percentage of arcs and nodes in congestion classes with $\alpha = 1$ and $\alpha = 0$

increase in walking time is minimized is very low, while it arises to almost 70% when total increase in walking time is neglected and all the focus is on minimizing congestion. As noted also above, just a small increase on the length of the allowed paths gives a high reduction of congestion.

4 Conclusions and Future Research

In this paper we presented a multi-objective model for the Gathering Avoiding Centralized Pedestrian Routing Problem. Its aim is to ensure pedestrians can walk in safety according to COVID-19 regulations, which request people to social distancing. Hence, the objectives were to minimize the total increase in time needed by the users to reach their destinations and to minimize the exceeding of capacity on the arcs and at nodes. Taking into account users' fairness, the eligible paths are chosen among the ones that do not exceed the length of the shortest path of more than a proper percentage. Because of the contrasting nature of these objectives, the problem is modeled as a multi-objective model.

The computational experiments were carried on by combining the two objectives linearly. The results are promising and show that just a small increase in the length of the paths that are taken into account can lead to a substantial reduction in congestion. Moreover, by opportunely setting parameter α it is possible to ensure the levels of unfairness among users is controlled.

Future researches may improve the solution method by exploring different methods for multi-objective models. Heuristic methods to handle very large instances could be explored too.

References

1. Angelelli, E., et al.: Proactive route guidance to avoid congestion. Transport. Res. B Methodol. **94**, 1–21 (2016)
2. Angelelli, E., et al.: Congestion avoiding heuristic path generation for the proactive route guidance. Comput. Oper. Res. **99**, 234–248 (2018)
3. Angelelli, E., et al.: A trade-off between average and maximum arc congestion minimization in traffic assignment with user constraints. Comput. Oper. Res. **110**, 88–100 (2019)
4. Angelelli, E., et al.: System optimal routing of traffic flows with user constraints using linear programming. Eur. J. Oper. Res. (In press, 2020)
5. Angelelli, E., et al.: Minimizing the total travel time with limited unfairness in traffic networks. Comput. Oper. Res., 123 (2020)
6. Avineri, E.: Social value orientation and the efficiency of traffic networks. Emerald (2009)
7. Chiandussi, G., et al.: Comparison of multi-objective optimization methodologies for engineering applications. Comput. Math. Appl. **63**, 912–942 (2012)
8. Dal Sasso, V., et al.: Incorporating Stakeholders' priorities and preferences in 4D trajectory optimization. Transport. Res. B Methodol. **117**, 594–609 (2018)
9. Fehr, E., et al.: The nature of human altruism. Nature, 425 (2003)
10. Jahn, O., et al.: System-optimal routing of traffic flows with user constraints in networks with congestion. Operations Research **53**, 600–616 (2005)
11. Mahmassani, H.S., et al.: On boundedly rational user equilibrium in transportation systems. Transportation Science **21**, 89–99 (1987)
12. Mahmassani, H.S., et al.: Network performance under system optimal and user equilibrium dynamic assignments: implications for advanced traveler information systems. Transport. Res. Rec. (1993)
13. Morandi, V.: Bridging the user equilibrium and the system optimum in static traffic assignment: how the cooperation among drivers can solve the congestion problem in city networks (2020). Available as working paper in https://valentinamorandi.it/research-outcomes/
14. Roughgarden, T., et al.: How bad is selfish routing? J. ACM (JACM), 49 (2002)
15. Schulz, A.S., et al.: Efficiency and fairness of system-optimal routing with user constraints. Networks **48**, 223–234 (2006)
16. Sheffi, Y.: Urban transportation network. Equilibrium Analysis with Mathematical Programming Methods. Prentice Hall (1985)
17. Van Essen, M., et al.: From user equilibrium to system optimum: a literature review on the role of travel information, bounded rationality and non-selfish behaviour at the network and individual levels. Transport Reviews **36**, 527–548 (2016)
18. Wardrop, J.G., et al.: Road paper. Some theoretical aspects of road traffic research. Proc. Inst. Civil Eng. **1**, 325–362 (1952)
19. Zhang, L.: Behavioral foundation of route choice and traffic assignment: comparison of principles of user equilibrium traffic assignment under different behavioral assumptions. Transport. Res. Rec. **254**, 1–10 (2011)

A MILP Formulation for the Reorganization of the Blood Supply Chain in Italian Regions

Antonio Diglio, Andrea Mancuso, Adriano Masone, Carmela Piccolo, and Claudio Sterle

Abstract Blood is a vital resource for human being since its unavailability may cause deaths and complications for patients. For this reason, in the last 20 years great interest has been addressed worldwide to the blood supply chain management, in terms of efficient and effective policy making and system design.

In answer to this raising need, the Italian Healthcare Ministry issued a decree aimed at improving the blood supply chain at regional level in terms of costs, accessibility and shortage, providing also several indications and restrictions to be accounted for. In this context, this chapter presents a mixed-integer linear programming formulation to determine the optimal location and the number of blood facilities at regional scale, with the aim of minimizing system costs meanwhile guaranteeing good standard service level requirements.

The proposed formulation tackles the problem in a multi-echelon and multi-objective perspective. It has been solved by Gurobi 9.0.1 solver and validated on real like test cases related to Campania and Puglia regions. Finally, the impacts of the different parameters involved in the formulation are analyzed in order to provide managerial insights to decision makers and healthcare stakeholders.

Keywords Blood Supply Chain · Facility Location · MILP formulation · Multi-objective

A. Diglio · C. Piccolo
Department of Industrial Engineering (DII), Università degli studi di Napoli Federico II, Napoli, Italy
e-mail: Antonio.Diglio@unina.it; Carmela.Piccolo@unina.it

A. Mancuso (✉) · A. Masone · C. Sterle
Department of Electrical Engineering and Information Technology (DIETI), Università degli studi di Napoli Federico II, Napoli, Italy
e-mail: Andrea.Mancuso@unina.it; Adriano.Masone@unina.it; Claudio.Sterle@unina.it

A. Masone et al. (eds.), *Optimization and Data Science: Trends and Applications*, AIRO Springer Series 6, https://doi.org/10.1007/978-3-030-86286-2_5

1 Introduction

Blood has several functions, including transporting oxygen and nutrients to the lungs and tissues [1]. When extracted from the human body, it becomes a perishable product that can be used for medical treatments, such as surgery, organ transplantation, and treatments for cancer and blood disorders. No substitute of human blood is currently available and, in many countries, blood itself is considered a scarce resource, whose waste is unacceptable as a harbinger of postponed surgery, untreated patients and deaths [2]. Voluntary donation is the main source of blood, suffering from some drawbacks such as the limited number of donors, test delays and high perishability [3]. Some blood substitutes, such as artificial blood or blood components with in vitro stem cells, may become alternatives only in the future [4].

In this context, the blood supply chain (BSC) management is a crucial activity. It is devoted to the management of the six main processes associated with blood flow: collection, testing, component processing, storage, distribution, and transfusion. These activities develop through five interacting levels to be opportunely coordinated: donors, mobile collection sites (CSs), blood centers (BCs), demand nodes (hospitals, clinics, or other transfusion points), and patients [5].

Optimal decisions making in the BSC is thus necessary to minimize the shortages, the wastes and to design a more efficient, effective and robust system.

In Italy the BSC is managed by the Transfusion System, which is a public body part of the Italian National Health Service (NHS) established in 1978. Unlike other countries, the Italian BSC is fully based on voluntary and unpaid donations, currently involving around 1.7 million donors, 1.3 million of which are periodic and over 370,000 are occasional first donors. On average, a blood donation occurs every 10 s, which enables approximately 1745 patients to be transfused and thousands of people to be treated with plasma-derived medicinal products per day [6].

In 2012, the Italian government, following the European Directive 2002/98/EC, issued a series of national guidelines for regional authorities related to the accreditation of transfusion units dedicated to the collection and processing of blood and its components [7]. In particular, some efficiency measures have been introduced, according to which a consolidation of processing activities (blood analysis, separation of blood components and transformation into plasma-derived products) must be performed at a limited number of Blood Centers (BCs). Based on this recommendation, a minimum of 40,000 units per year and for each BC should be guaranteed. This target is significantly higher than the average BC productivity in all the Italian regions. Indeed, the average BC productivity in Italy is of 9981 units per year and, in the regions objective of our experimentation, namely Campania and Puglia, this value is of 6890 and 7682 units per year, respectively [6].

Therefore, a reorganization of Italian regional BSC is required with the aim of rationalizing its structure and guaranteeing the regional self-sufficiency. The regional authorities recognize two main intervention strategies to increase the BC productivity until the imposed threshold value: decreasing the number of available BCs, choosing the ones to be left open among the already existing ones; increasing the blood units assigned to each BC by employing mobile units to reach isolated blood donation points. In addition to these strategies, as proposed in [8], we consider also the possibility of performing a downgrade of the existing BCs to the so called Blood Stations (BS), where only collection activities are carried out. The blood collected at the BSs is then transferred to the BCs for the processing. Since the propensity to donation is highly affected by the distance between the donors and the collection facility, the introduction of such BSs should allow to further increase the collected blood units per year. In the following we refer BC and BS as Blood Facilities (BF).

As widely known, Operations Research methodologies represent a valuable solution tool for decision problems arising in healthcare planning and management. As a matter of fact, optimization problems in healthcare have received considerable attention for over three decades, as shown in [9–11]. In particular, the literature concerning BSC management is very ample and deals with many different strategic, tactical and operational problems, differing for the time horizon, the nature of the available data, the envisaged processes and the structure (number of levels) of the system.

This work focuses on a strategic single period decision problem for the reorganization of a multi-echelon BSC system. This problem falls within the classes of covering and facility location problems (FLP). For the sake of the brevity, in the following, we just give a brief overview of the literature dealing with strategic BSC management problems coherent with the problem tackled in this work.

Several contributions deal with BSC management problems arising in specific regions. Indeed, in [12] the authors tackle the problem of allocating health resources throughout the area of Burgos, in the north of Spain. It is modeled and solved as a variant of the p-center and maximal covering problems. In [13] a FLP related to the blood services of Turkish Red Crescent at a regional scale is addressed. The authors considered the problem as a hierarchical system in which we define two echelons in the hierarchy of services. In [14] a capacitated maximal covering problem is proposed to locate healthcare facilities in one of the districts in the state of Selangor, Malaysia. Recently, in [15] the authors analyzed a strategic problem concerning the territorial reorganization of regional blood management systems envisaging two kind of blood facilities. The study was aimed at closing/reconverting a set of operating BFs, already located in a region, to find a good trade-off solution to attract donors and contain the management costs. The model was tested on the real case study of Campania Region.

Other contributions, instead deal with more general problems and introduce new requirements related to accessibility and quality of service. In [16] the authors investigate the problem of locating several different facilities (e.g., hospitals, universities, police workstations) in an urban area. Different optimization models are proposed in dependence of the specific kind of facility and targets. The models are then validated on several test cases so showing that they can represent a valuable support tool for quantitative decision making. The concept of accessibility is introduced in [15], where the authors provide a review of the main methodological advancements dealing with the inequality issue in healthcare services. Various methods have been proposed to measure healthcare accessibility, accounting for both spatial and nonspatial factors. In [17] the blood collection and distribution network design problem in response to the growing need for effective and efficient management of blood systems is addressed. A mixed integer linear programming (MILP) model is provided to design a cost-efficient network in terms of blood facilities. In [18] the authors demonstrate the possibilities of improving the supply of blood products while also reducing transportation costs within available budgets. Integer linear programming (ILP) is used to model this location–allocation problem based on contrasting objectives of improving supply while reducing transportation costs.

Starting from the analyzed literature, in this chapter we extend the work presented in [8] introducing new set of requirements and multiple objectives which take into account accessibility and quality service level, coherently with what has been done in the other cited contributions. In particular, the proposed reorganization of the regional BSCs configures a multi-echelon facility location problem where the number and the position of the BFs (BCs and BSs) has to be determined keeping into account also the usage of collection mobile units. This chapter proposes a MILP formulation for the problem, taking into account operational constraints and several side objectives of the system. The multi-objective aspect is treated in a soft constraint fashion. The proposed formulation is experienced and validated on real data sets referred to the Campania and Puglia regions, showing the benefits of such reorganization and providing also several managerial insights which can be replicated in other contexts.

The remainder of the chapter is structured as follows. A formal description of the problem and the mathematical model are given in Sect. 2. Section 3 is devoted to the presentation of our case study and to the discussion of the results. Finally, Sect. 4 provides the conclusions.

2 MILP Model for the Reorganization of a Regional BSC

The reorganization of the regional BSC involves two strategic decisions, one related to the employed facilities and the other related to the donor assignments. In particular, concerning the decisions regarding the blood facilities, each existing structure can be:

- Held open as BC;
- Downgraded to be a BS, disabling blood processing and allowing only the collection activity, which obviously implies lower running costs;
- Closed and no longer used.

In terms of assignment, we have to decide the one-to-one assignment of each donor to the available BS, BC or mobile unit.

To model the problem, it is made reference to a localization space corresponding to a generic region associated with an annual demand D for blood to be collected.

It is assumed that there is a set J of BCs located in the area to carry out both blood collection and processing activities and a set I of discrete nodes, where potential donors are concentrated. On the basis of these sets, we define d_{ij} as the distance from a node i, $i \in I$, and a facility j, $j \in J$, and $d_{jj'}$ as the distance between the pair of facilities (j, j'), $j, j' \in J$, respectively. More precisely, d_{ij} represents the distance each donor is willing to travel and it has to be lower than or equal to a maximum radius r. The blood collected in each BS must be transported to an active BC in order to be processed. Due to some technical constraints, related to the phenomenon of blood degradation, it is assumed that this allocation is possible if the distance $d_{jj'}$ is equal to or lower than a maximum radius d_{max}. Furthermore, it is assumed that each BS can be assigned to a single BC, thus avoiding the possibility of dividing its blood collection among different BCs; this hypothesis is justified by the fact that, from a managerial point of view, it is desirable to define a group of BSs under the competence of a single BC in order to better coordinate the associated flows of materials and information. Based on previous notation, three more additional sets are needed to model the problem: N_i, set of nodes j, $j \in J$, whose distance from the i, $i \in I$, is less than or equal to r; M_j, set of nodes j', $j' \in J$, whose distance from j, $j \in J$ is less than or equal to d_{max}; O_i, set of nodes j, $j \in J$, whose distance from the i, $i \in I$, is less than or equal to d_{max}.

Each BS is characterized by a collection capacity C, which indicates the maximum number of units of whole blood it is able to collect each year. It is also assumed that in each node i, $i \in I$, we have quantity of potential donors equal to a_i $(\alpha * p_i)$, where α is the donation rate and p_i the population of the node i, respectively. Facilities carrying out processing activities are required to achieve a minimum productivity target P_{min} in terms of whole blood produced per year, according to the guidelines issued by the national government. Assuming that the regional transfusion system does not meet this requirement in the current configuration, the reorganization is aimed at consolidating the processing activities in a smaller number of facilities while maintaining a capillary collection in the area under investigation. To this aim, it is assumed that the region is equipped with a set of mobile units, in a number equal to n, which, unlike the fixed structures, are able to collect the blood units at the donors location. Each mobile unit has a maximum annual collection capacity t, and that it transports the collected units to a BC for subsequent processing steps. As previously explained, it is assumed that the transfer of the blood units collected in a generic node is carried out towards a single BC located within the maximum radius d_{max}.

Therefore, the problem consists in choosing the right combination of BC and BS among the existing BFs, with the aim of minimizing the associated costs, that is, the costs necessary to move the blood units collected by the BS and mobile units to the BCs (these costs are assumed to be proportional to the amount of blood collected and the distances traveled), while satisfying a series of requirements in terms of total blood collected (self-sufficiency goal) and productivity for each BC (efficiency goal, i.e., reaching the minimum productivity avoiding the BF capacity overrunning). These requirements are to be understood as soft requirements, since they can be violated by paying a penalty cost, proportional to a parameter λ in the objective function. According to these strategic decisions, the following decision variables are defined:

Node variables (binary):

$y_j^{(s)}$ equal to 1 if a BS is active in node j, $j \in J$, in the final configuration, 0 otherwise;

$y_j^{(c)}$ equal to 1 if a BC is active in node j, $j \in J$, in the final configuration, 0 otherwise;

Assignment variables (binary):

x_{ij} equal to 1 if all the potential donors of node i, $i \in I$, are assigned to facility in $j, j \in J$, 0 otherwise;

$q_{jj'}$ equal to 1 if the amount of blood collected from the structure in $j, j \in J$, is processed by the structure in $j', j' \in J$, 0 otherwise;

z_{ij} equal to 1 if the potential donors of the node i, $i \in I$, are assigned, through the use of mobile units, to the BC in $j, j \in J$, 0 otherwise;

$t_{ijj'}$ equal to 1 if the potential donors of the node i, $i \in I$, initially assigned to the localized structure in $j, j \in J$, are reallocated to the BC in $j', j' \in J$, 0 otherwise;

$w_{ij'j}$ equal to 1 if both the variables $x_{ij'}$ and $q_{j'j}$ are equal to 1, 0 otherwise. These variables are needed to linearize the product $x_{ij'}\, q_{j'j}$.

Penalty variables (non-negative):

φ_j shortage of processed blood with respect to minimum productivity requirement P_{min} of a BC in $j, j \in J$;

ψ_j surplus with respect to the satisfaction of the maximum collection capacity requirement C of the structure in $j, j \in J$;

δ scarcity in terms of fulfilment of the regional self-sufficiency requirement D.

In Fig. 1, we provide a graphical representation of the multi-echelon system under investigation, with the values assumed by the related decision variables and distance parameters.

$$\text{Min } z = \sum_{i\in I}\sum_{j\in J}\sum_{j'\in J} a_i d_{jj'} w_{ijj'} + \sum_{i\in I}\sum_{j\in J} a_i d_{ij} z_{ij} + \sum_{i\in I}\sum_{j\in J}\sum_{j'\in J} a_i d_{jj'} t_{ijj'} + \sum_{j\in J} \lambda \left(\varphi_j + \psi_j + \delta\right) \quad (1)$$

s. t.

$$y_j^{(s)} + y_j^{(c)} \le 1 \qquad \forall j \in J \quad (2)$$

$$x_{ij} \le y_j^{(s)} + y_j^{(c)} \qquad \forall i \in I, \forall j \in N_i \quad (3)$$

$$\sum_{j\in(J-N_i)} x_{ij} = 0 \qquad \forall i \in I \quad (4)$$

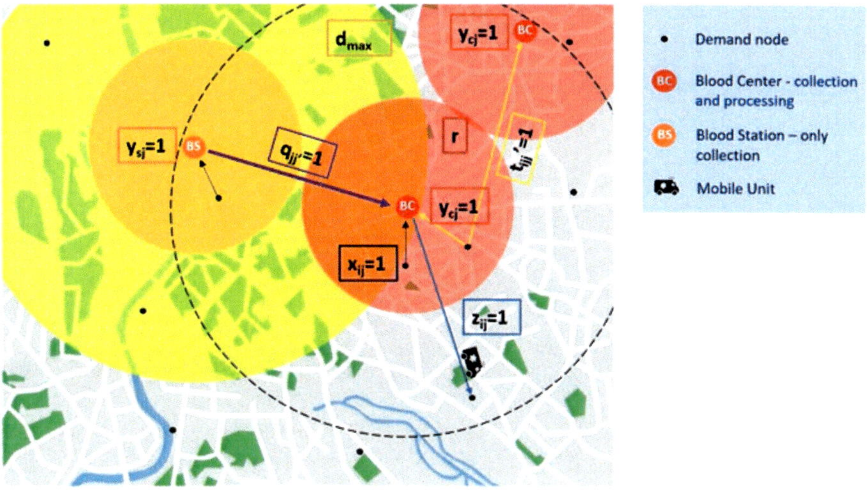

Fig. 1 Representation of the problem setting, parameters and variables for a region under investigation

$$| N_i | \sum_{j \in N_i} x_{ij} \geq \sum_{j \in N_i} \left(y_j^{(s)} + y_j^{(c)} \right) \qquad \forall i \in I \qquad (5)$$

$$\sum_{j' \in N_i} d_{ij'} x_{ij'} + \left(F - d_{ij} \right) \left(y_j^{(s)} + y_j^{(c)} \right) \leq F \qquad \forall i \in I, \forall j \in N_i \qquad (6)$$

$$q_{jj'} \leq y_{j'}^{(c)} \qquad \forall j \in J, \forall j' \in J \qquad (7)$$

$$\sum_{j' \in M_j} q_{jj'} = y_j^{(s)} + y_j^{(c)} \qquad \forall j \in J \qquad (8)$$

$$\sum_{j' \in (J - M_j)} q_{jj'} = 0 \qquad \forall j \in J \qquad (9)$$

$$q_{jj} \geq y_j^{(c)} \qquad \forall j \in J \qquad (10)$$

$$\sum_{j \in J} \left(x_{ij} + z_{ij} \right) \leq 1 \qquad \forall i \in I \qquad (11)$$

$$z_{ij} \leq y_j^{(c)} \qquad \forall i \in I, \forall j \in O_i \qquad (12)$$

$$\sum_{j \in (J - O_i)} z_{ij} = 0 \qquad \forall i \in I \qquad (13)$$

$$\sum_{j \in J} z_{ij} \leq 1 \qquad \forall i \in I \qquad (14)$$

$$\sum_{j' \in J} t_{ijj'} \leq x_{ij} \qquad \forall i \in I, \forall j \in J \qquad (15)$$

$$t_{ijj'} \le y_{j'}^{(c)} \qquad \forall i \in I, \forall j, j' \in J \qquad (16)$$

$$\sum_{i\in I}\sum_{j\in J} a_i x_{ij} + \sum_{i\in I}\sum_{j\in J} a_i z_{ij} + \delta \ge D \qquad (17)$$

$$\sum_{i\in I} a_i x_{ij} - \sum_{i\in I}\sum_{j'\in J} a_i t_{ijj'} - \psi_j \le C \qquad \forall j \in J \qquad (18)$$

$$\begin{aligned}&\sum_{i\in I}\sum_{j'\in J} a_i \left(w_{ij'j} - t_{ij'j}\right) - \sum_{i\in I}\sum_{j'\neq j\in J} a_i t_{ijj'} + \\ &\sum_{i\in I}\sum_{j'\in J} a_i t_{ij'j} + \sum_{i\in I} a_i z_{ij} + \varphi_j \ge P_{min}\, y_j^{(c)}\end{aligned} \qquad \forall j \in J \qquad (19)$$

$$\sum_{i\in I}\sum_{j\in J} a_i z_{ij} + \sum_{i\in I}\sum_{j\in J}\sum_{j'\in J} a_i t_{ijj'} \le tn \qquad (20)$$

$$w_{ijj'} \le x_{ij} \qquad \forall i \in I, \forall j, j' \in J \qquad (21)$$

$$w_{ijj'} \le q_{jj'} \qquad \forall i \in I, \forall j, j' \in J \qquad (22)$$

$$w_{ijj'} \ge x_{ij} + q_{jj'} - 1 \qquad \forall i \in I, \forall j, j' \in J \qquad (23)$$

$$\begin{aligned}&y_j^{(s)}, y_j^{(c)}, x_{ij}, q_{jj'}, w_{ijj'}, z_{ij}, t_{ijj'} \in \{0, 1\} && \forall i \in I, \forall j, j' \in J \qquad (24)\\ &\varphi_j, \psi_j \ge 0 && \forall j \in J\\ &\delta \ge 0\end{aligned}$$

The objective function (1) minimizes four components. The first two components concern the transportation costs from BS to BC and from donors to BC (using mobile unit), respectively. The third component considers the cost for the reallocation of a donor node from the nearest BC to a different one in case of capacity overrun. Finally, the fourth component includes a penalty cost linked to the non-compliance with the requirements of efficiency, self-sufficiency and the exceeding of the maximum collection capacity. The constraints (2) ensure that a single type of BF can be located in each node $j, j \in J$. Constraints (3) are consistency constraints between assignment and node variables. Constraints (4) avoid the allocation of any node $i, i \in I$, to BF $j, j \in J$, furthest away from the coverage radius r. The constraints (5) impose that each node $i, i \in I$, must be assigned to at least one structure within its coverage radius. The constraints (6) require the node $i, i \in I$, to be allocated, within the radius r, to the nearest active structure. As a matter of fact, by setting $F = max_{i \in I, j \in J}\{d_{ij}\}$ the conditions become redundant for all those nodes where no structure is found while, in relation to the active BF, they assign each node $i, i \in I$, to node $j', j' \in J$, at the minimum distance. The constraints (7) ensure that the node j, $j \in J$, can be assigned only to structures $j', j' \in J$, where a BC is open. Constraints (8) dictate that any node $j, j \in J$, in which a BF is active in the final configuration must be assigned to a single structure within the maximum allowable distance d_{max}. The constraints (9) avoid the allocation of any BS j, $j \in J$, to the BC j', $j' \in J$, farthest from the coverage radius d_{max}. Constraints (10) ensure that each BC is assigned to itself. The constraints (11) ensure that each node i, $i \in I$, can be allocated to a single structure j, $j \in J$, located within the radius r or less. In other words, these constraints establish that the potential donors offered by each node can be collected by direct transfer to an active BF or using mobile units. The constraints (12) impose that donors at nodes i, $i \in I$, allocated through mobile units and not covered by any active BF within the radius r, can be conferred for processing to an active BC j, $j \in J$, located within the covering radius d_{max} from node i, $i \in I$. This condition is strengthened by the constraints (13) which avoid the allocation to BC j, $j \in J$, placed further than the radius d_{max} from node i, $i \in I$. Furthermore, the constraints (14) ensure that donors in each node i, $i \in I$, allocated through mobile units can be assigned to only one BC. The constraints (15) establish that the reallocation of a node i, $i \in I$, or the allocation through mobile units of the donors offered by a node i, $i \in I$, located at a radius less than r from at least one active center can be only carried out if this node has already been assigned to a structure j, $j \in J$. Constraints (16) impose that these reallocations can occur only in correspondence with active BCs in j', $j' \in J$. Constraint (17) ensures that the amount of whole blood collected is at least equal to the total regional blood demand D (self-sufficiency goal). Constraints (18) define an upper limit to the capacity for the collection service at each facility (efficiency goal). The constraints (19) require that the total amount of whole blood processed in each BC, collected by the latter, by the BS and by the mobile units assigned to it, exceeds the minimum productivity target P_{min}(efficiency goal). Constraint (20) ensure that the total amount of blood collected through the mobile units does not exceed the maximum available capacity (tn). Constraints (21), (22) and (23) ensure that the new binary variables $w_{ijj'}$, introduced to linearize the proposed model, are equal to 1 if donors from node i, $i \in I$ are allocated to structure j', $j' \in J$ and worked in j, $j \in J$. Finally, the constraints (24) define the nature of the decision variables introduced.

It is useful to clarify that the decision variable $t_{ijj'}$ has a twofold utility: on the one hand it allows to extend the coverage radius by assigning nodes out of the maximum coverage radii r and d_{max}, when the model requires it; on the other hand it allows the model to choose whether to reassign the nodes or pay the penalty cost for exceeding the maximum capacity C.

3 Application of the Model to the Case of the BSC of the Campania and Puglia Regions

In this section we validate the proposed MILP formulation on two test cases related to two Italian regions, namely Campania and Puglia. In particular, Sect. 3.1 describes the problem setting and used parameters; Sect. 3.2 is devoted to the presentation and discussion of the obtained experimental results.

3.1 Test Case Description

Among all the Italian regions the chapter refers to Campania and Puglia for two main reasons. The first reason is of geographical/topological nature, since Campania has a compact shape while Puglia has an elongated one. The second reason is the presence in Campania of the city of Naples acting as a very strong center of gravity for the BCs, since its number of inhabitants and population density is largely higher than the other provinces of the same region. These two different regions are considered to investigate if the aforementioned conditions have a significant effect on the solution returned by the proposed model. For both areas under investigation a discretization based on the pre-exiting territorial division of the region in municipalities is adopted. Donations are considered to be proportional to population in each municipality by a coefficient α (donation rate) equal to the percentage of population that donates blood according to historical data. It has been assumed that every person donates once every year and that coefficient α is uniformly distributed within the region. According to historical data provided by the regional authority, it has been observed that, on average, in Italy almost 35 donations every 1000 inhabitants have been carried out in the last 5 years ($\alpha = 0.0350$). Campania is currently characterized by the presence of 22 BCs, while Puglia has 21 opened ones. Both the regions have an average level of productivity significantly lower than the minimum threshold P_{min} imposed by EU and national regulation. Furthermore, as far as the Campania region is considered, due to the presence of the municipality of Naples, characterized by a largely higher population level than the other cities of the region (almost 962,000 inhabitants) and by the presence of nine BCs, the municipal territory has been divided into ten sub-areas in order to obtain experiments of greater accuracy. In addition, the λ penalty value, based on a calibration stage, has been set to a very large value ($\lambda = 10^5$), since not achieving the requirements of self-sufficiency and efficiency is allowed but it should be strongly avoided.

The three calibration parameters (r, d_{max}, α) have been varied to assess the stability of the optimal solution.

In particular:

- The coverage radius r of a center reproduces the willingness of people to travel longer distances following regional or national awareness-raising campaigns of sensitization ($r = 20$ and 30 km);
- The coverage radius d_{max} evaluates the effects of the different blood transport strategies collected by BS to the assigned BC and mobile units to the centers ($d_{max} = 40$, 50 and 60 km);
- The donation rate α simulates the effects of potential regional campaigns of awareness aimed at increasing the propensity of the population to donate blood, or the decrease of this rate from its present value to simulate a stressful situation for regional BMS ($\alpha = 0.030$, 0.035 and 0.040).

All the experiments carried out will be evaluated in terms of open BC and BS, objective function value and average accessibility value ($\overline{a}$), defined as the sum of minimum Euclidian distances of a node i where the donators are concentrated to an open BC or BS in the considered scenario from $\left(\overline{a} = \sum_{i \in I} \frac{d^*_{ij}}{|I|}\right)$ [19].

3.2 Experimental Results

This section will provide extensive results for both the regions. Furthermore, graphical representations of several experienced scenarios are provided.

The test problems, generated for each combination of the above parameters, were solved using Gurobi 9.0.1 solver on an Intel Core i7 (with 2.60 GigaHertz and 16.00 GigaBytes of RAM).

Table 1 shows the value of objective function (column 2), the average accessibility (column 3) and the number of BCs and BSs opened in the scenario (column 7, 8) for every possible combination of the calibration parameters (column 4, 5, 6) considered in the study for the Campania region (we will refer to every possible configuration of the parameter as an instance, indicated by letter C followed by the number of the scenario, note that the C19 represent the value for the as is scenario). Table 2 is referred to the Puglia region (we will refer to every possible configuration

Table 1 Results for the Campania region varying the $\boldsymbol{\alpha}$,$\boldsymbol{d_{max}}$ and $\overline{a}$ parameters

Scenario	Z	$\overline{a}$	d_{max}	α	r	BCs	BSs	Time (s)
C1	1.73 E+09	25.94	40	0.035	30	1	3	917
C2	1.23 E+09	24.93	50	0.035	30	2	3	956
C3	1.23 E+09	24.93	60	0.035	30	2	3	902
C4	8.02 E+08	32.99	40	0.04	30	1	3	481
C5	1.20 E+09	27.63	40	0.03	30	2	6	998
C6	8.06 E+08	26.62	50	0.03	30	2	3	1522
C7	7.65 E+08	27.19	50	0.04	30	2	2	632
C8	8.05 E+08	26.62	60	0.03	30	2	3	1270
C9	7.64 E+08	27.19	60	0.04	30	2	2	654
C10	9.14 E+08	29.77	40	0.035	20	2	3	342
C11	2.61 E+09	27.55	40	0.03	20	2	6	550
C12	2.68 E+09	28.40	40	0.04	20	1	4	307
C13	9.14 E+08	26.83	50	0.035	20	2	4	598
C14	2.19 E+09	24.59	50	0.03	20	1	6	519
C15	2.93 E+07	25.55	50	0.04	20	1	4	565
C16	9.14 E+08	26.83	60	0.035	20	2	5	541
C17	2.19 E+09	24.59	60	0.03	20	1	6	491
C18	2.85 E+07	25.55	60	0.04	20	2	4	254
C19 (as-is)	6.83 E+11	16.48	40	0.035	30	22	0	23

Table 2 Results for the Puglia region varying the α,d_{max} and $\overline{a}$ parameters

Scenario	Z	$\overline{a}$	d_{max}	α	r	BCs	BSs	Time (s)
P1	3.28 E+06	46.08	40	0.035	30	2	2	52
P2	2.09 E+06	39.33	50	0.035	30	2	4	55
P3	1.75 E+06	38.78	60	0.035	30	2	4	61
P4	4.27 E+05	53.39	40	0.04	30	2	2	24
P5	8.07 E+08	32.33	40	0.03	30	2	2	49
P6	5.41 E+06	29.47	50	0.03	30	2	4	63
P7	4.27 E+05	53.39	50	0.04	30	2	4	46
P8	4.81 E+06	35.91	60	0.03	30	2	4	70
P9	4.27 E+05	53.39	60	0.04	30	2	4	54
P10	2.79 E+08	32.45	40	0.035	20	2	4	13
P11	1.80 E+09	26.94	40	0.03	20	2	4	12
P12	2.71 E+06	46.06	40	0.04	20	2	4	14
P13	6.41 E+06	41.09	50	0.035	20	2	4	10
P14	1.31 E+09	38.81	50	0.03	20	2	4	16
P15	2.71 E+06	46.06	50	0.04	20	2	4	12
P16	5.02 E+06	36.28	60	0.035	20	2	4	11
P17	7.65 E+08	33.5	60	0.03	20	2	4	18
P18	1.14 E+06	38.78	60	0.04	20	2	4	11
P19 (as-is)	7.28 E+14	17.75	40	0.035	30	21	0	7

of the parameter as an instance, indicated as P followed by the number of the scenario, note that the P19 represent the value for the as is scenario). As clearly shown in Tables 1 and 2 and Figs. 2, 3, and 4, two main trends emerged in both Campania and Puglia. First of all, even in the worst case scenario the objective function is significantly lower than the one in the "as-is" scenario. Moreover, solutions move from the 22 open facilities for the Campania region (all BC in the current configuration) to 8 centers (2 BC and 6 BS) in the worst case scenario. A similar trend is observed for the Puglia region moving from the actual 21 open centers (all BC) to 6 open centers (2 BC and 4 BS) in the scenario presenting the worst values. Concerning computation time, as can be seen in the last column of Tables 1 and 2, almost all cases have been solved within 900 s, which is acceptable for strategic long-term decision problems. In addition, the experiments show that the objective function has a decreasing trend as calibration parameters (α and r) increase, which could be an indicator of how certain population awareness campaigns can make the model more efficient regardless of the management policies implemented, because they affect the maximum distance that donors are willing to cover (r) and the percentage of people making donations (α). Finally, collected results indicates that policies for optimal transport of blood products would allow the d_{max} radius to be increased.

Among the analyzed scenarios, four cases of interest deserve to be further discussed:

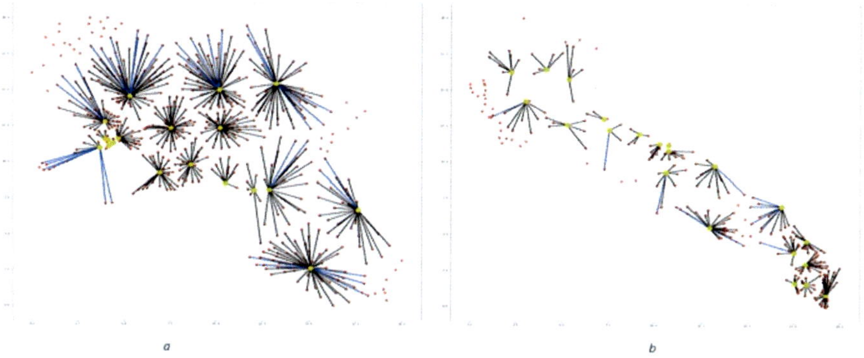

Fig. 2 Representation of the solutions of the "as-is" scenarios: (**a**) solution for Campania region; (**b**) solution for Puglia region

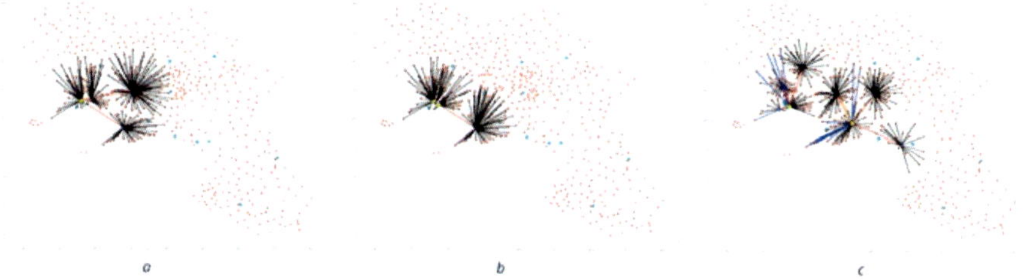

Fig. 3 Representations of the solutions obtained for Campania Region: (**a**) solution with "as-is" parameters; (**b**) solution of the $\overline{a}$ worst case scenario; (**c**) solution of the z worst case scenario

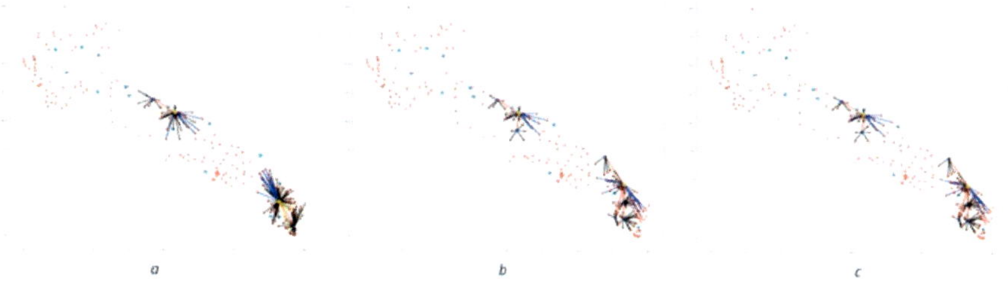

Fig. 4 Representations of the solutions obtained for Puglia Region: (**a**) solution with "as-is" parameters; (**b**) solution of the $\overline{a}$ worst case scenario ; (**c**) solution of the z worst case scenario

1. The "as-is" scenario, in which we constrained the model to let open all the BCs to simulate the situation currently running in Italy, using as combination of parameters $\alpha = 0.035$, $r = 30$, $d_{max} = 40$, which are given by the regional authority, in order to have a benchmark to compare the solutions proposed by our experimentation (Fig. 2a for Campania, Fig. 2b for Puglia);

2. The scenario running the same configuration of parameters as in the "as-is" scenario, i.e., $\alpha = 0.035$, $r = 30$, $d_{max} = 40$, but letting the model decide which facility to left open as a BC, which one to downgrade (BS) and which one to close (Fig. 3a for Campania, Fig. 4a for Puglia);
3. The worst case of average accessibility which corresponds to the following parameter values: $\alpha = 0.04$, $r = 30$, $d_{max} = 40$ for Campania; $\alpha = 0.03$, $r = 30$, $d_{max} = 40$ for Puglia (Fig. 3b for Campania, Fig. 4b for Puglia);
4. The worst case of the objective function value which corresponds to the following parameter values:$\alpha = 0.03$, $r = 20$, $d_{max} = 40$ for both regions (Fig. 3c for Campania; Fig. 4c for Puglia).

For all this cases the solutions are drawn using the following notation:

- The red dots represent the centroids of each municipality in which the population of the entire municipality is concentrated;
- The yellow dots represent the open BCs in the configuration considered;
- The blue squares represent BSs in the configuration considered;
- The black arcs indicate which BC, open in the configuration considered, is associated with the demand present in each centroid;
- The blue arcs, when present, indicate that the demand present in the node is associated with BC through the use of mobile units;
- The yellow links, when present, indicate that the demand for the node, initially allocated to a facility, is reallocated to another facility through the use of mobile units.

We can easily note that the *as-is* scenarios of both regions employ a high number of BFs. On the contrary, the productivity threshold of the other scenarios decreases the number of facilities of one order of magnitude. Such decrease reflects in low accessibility values and few facilities placed in more densely populated area, so pushing towards a high exploitation of mobile units.

4 Conclusions

This work analyzes the strategic problem concerning the reorganization of regional blood management systems. It is aimed at closing or reconverting the facilities already operating in a region and find a good compromise solution between the objective of attracting donors (objective of self-sufficiency) and the need of reducing the management costs (objective of efficiency). To this end, an optimization model is proposed and developed, taking inspiration from the literature on multi-echelon facility location. The main objective of this work is to provide an insight of the impact of possible strategies on the regional BSC in order to support the competent authorities in its reorganization, and to evaluate the sensitivity of the proposed solutions to the variation of several parameters, in order to assess the impact of certain policies on the proposed strategies. The model was tested on two real case for the Campania and Puglia regions generating several scenarios. Varying some key parameters to simulate a wide range of real situations allowed to investigate the model robustness to the changing conditions, which is a very important factor when locating healthcare facilities. Furthermore, a possible future development of the problem is represented by the introduction of stochasticity in the problem, in line with what is happening in the literature [20–22]. In this context, the model should reallocate part of the donors to other centers to avoid stress scenarios caused by unforeseen situations.

References

1. American Society of Hematology. Blood Basics (2017)
2. World Health Organization, 10 Facts on Blood Transfusion (2017)
3. American Red Cross. Blood Components (2017)
4. Jansman, M.M., Hosta-Rigau, L.: Recent and prominent examples of nano and microarchitectures as hemoglobin-based oxygen carriers. Adv. Colloid Interface Sci. **260**, 65–84 (2018)
5. Pirabán, A., Guerrero, W.J., Labadie, N.: Survey on blood supply chain management: models and methods. Comput. Oper. Res. **112** (2019)
6. Italian National Health System. www.salute.gov.it/ (2014)
7. Agreement CSR Guidelines for the accreditation of blood transfusion services and units blood collection and blood components. Offic. J. Ital. Republic 149/2012 (2012)
8. Bruno, G., Diglio, A., Piccolo, C., Cannavacciuolo, L.: Territorial reorganization of regional blood management systems: evidences from an Italian case study. Omega. **89**, 54–70 (2018)
9. Beliën, J., Forcé, H.: Supply chain management of blood products: a literature review. Eur. J. Oper. Res. **217**(1), 1–16 (2012)
10. Rais, A., Viana, A.: Operations research in healthcare: a survey. Int. Trans. Oper. Res. **18**, 1–31 (2010)
11. Brailsford, A., Vissers, A.: OR in healthcare: a European perspective. Eur. J. Oper. Res. **212**, 223–234 (2011)
12. Pacheco, J.A., Casado, S.: Solving two location models with few facilities by using a hybrid heuristic: a real health resources case. Comput. Oper. Res. **32**, 3075–3091 (2005)
13. Sahin, G., Süralb, H., Meralb, S.: Locational analysis for regionalization of Turkish Red Crescent blood services. Comput. Oper. Res. **34**, 692–704 (2007)
14. Shariff, S., Hasnah, M.N., Omar, M.: Location allocation modeling for healthcare facility planning in Malaysia. Comput. Ind. Eng. **62**, 1000–1100 (2012)
15. Wang, F.: Measurement, optimization, and impact of health care accessibility: a methodological review. Ann. Assoc. Am. Geogr. **102**(5), 1104–1112 (2012)
16. Zhang H., Wang Z., Zhou W., Liu, X., 3, Yan, D.: Research and Application of Weber Facility Location (2006)
17. Zahiri, B., Torabi, S.A., Mousazadeh, M., Mansouri, S.A.: Blood collection management: methodology and application. Appl. Math. Model. **39**(23-24), 7680–7696 (2015)
18. Chaiwuttisak, P., Smith, H., Wu, Y., Potts, C., Sakuldamrongpanich, T., Pathomsiri, S.: Location of low-cost blood collection and distribution centres in Thailand. Oper. Res. Health Care. **9**, 7–15 (2016)
19. Bruno, G., Cavola, M., Diglio, A., Piccolo, C.: Improving spatial accessibility to regional health systems through facility capacity management. Socio Econ. Plan. Sci. **71** (2020)
20. Attari, M.Y.N., Aghaie, P.A.A., Niaki, S.T.A.: A bi-objective robust optimization model for a blood collection and testing problem: an accelerated stochastic Benders decomposition. Ann. Oper. Res. (2018). https://doi.org/10.1007/s10479-018-3059-9

21. Govindan, K., Fattahi, M., Keyvanshokooh, E.: Supply chain network design under uncertainty: a comprehensive review and future research directions. Eur. J. Oper. Res. **263**(1), 108–141 (2017)
22. Zhairi B., Mousazadeh M., Bozorgi-Amiri A.: A robust stochastic programming approach for blood collection and distribution network design. Int. J. Res. Ind. Eng. **3**(2), 1–11 (2014)

Part III
Logistics

Instance Generation Framework for Green Vehicle Routing

Matheus Diógenes Andrade and Fábio Luiz Usberti

Abstract This paper proposes a framework for generating relevant sets of instances for Green-Vehicle Routing Problems (G-VRP). In the G-VRP, electric vehicles with limited autonomy can recharge at Alternative Fuel Stations (AFSs) to keep visiting customers. To the best of our knowledge the G-VRP scientific literature accounts with only two sets of instances. Our instance generation framework is based on solving a maximum leaf spanning tree problem to address the location of AFSs, and it guarantees that the generated instances are feasible (as opposed to the procedure previously proposed). Two G-VRP variants are considered, (i) where consecutive AFSs visits are not allowed, and (ii) where consecutive AFSs visits are allowed. The results are analyzed and discussed, and conclusions on the benefits of the contributions are presented.

Keywords Green vehicle routing problem · Maximum leaf spanning tree problem · Instance generation · Green logistics

1 Introduction

The Vehicle Routing Problem (VRP) has been extensively studied in the literature, and many variants covering additional constraints related to scheduling, budget, fleet heterogeneity, and others are considered [24]. Another research field that has gained traction in the past decades is green logistics, which emerged like a

Supported by FAPESP (proc. 2015/11937-9, and 2018/25950-5), and CNPq (proc. 134616/2018-9, 314384/2018-9, and 435520/2018-0).

M. D. Andrade (✉) · F. L. Usberti (✉)
Institute of Computing, UNICAMP University of Campinas - Av. Albert Einstein, Campinas, Brazil
e-mail: fusberti@ic.unicamp.br

A. Masone et al. (eds.), *Optimization and Data Science: Trends and Applications*, AIRO Springer Series 6, https://doi.org/10.1007/978-3-030-86286-2_6

discipline focused on the planning and operation of logistic systems with high energy efficiency and low levels of carbon emission [20].

The environmental, geopolitical, and financial costs of the world's dependence on gasoline-powered vehicles are becoming increasingly worrisome [19]. The transport sector is responsible for high levels of CO_2 emissions. For example, the transport sector in 2018 accounted for 28.2% of CO_2 emission in the United States [14], and in 2014 it accounted for 20.45% of CO_2 emission worldwide [23]. Recognizing the environmental impacts, logistic companies are making their operations greener [17]. Furthermore, technological breakthroughs have enabled the development of *clean* motor vehicles, that do not emit harmful substances in their exhaust fumes and which create less noise without causing other types of pollution [12]. The adoption of Alternative Fuel Vehicles (AFVs) has been regarded as one of the most promising strategies to address the issues of fossil fuel dependence, air quality, and, more recently, climate change [26]. However, there are still several major barriers to broaden the adoption of AFVs. One of them is the maximum driving range that may be insufficient to complete some delivery tours [17], forcing the AFVs to stop in Alternative Fuel Stations (AFSs) in order to refuel. Thus, efficient methods that locate AFSs are essential to guarantee the routing decision feasibility. Locating AFSs to serve the maximum demand possible is the key to the early efforts in the design of new infrastructures composed of AFVs [19].

Among VRP variants dealing with green logistics, there is the Green Vehicle Routing Problem (G-VRP) [3]. The G-VRP has additional challenges associated with operating a fleet of AFV and incorporating stops at AFS [15]. One of the biggest challenges modeled by G-VRP is the reduced number of AFSs and the low autonomy of AFVs.

The current G-VRP literature lacks a more general set of instances (as far as we know, accounts for two sets [1, 15]) as well as a customizable framework to generate relevant sets of instances. In this paper, we will introduce such a framework to generate feasible G-VRP instances by solving a variant of the Maximum Leaf Spanning Tree (MLST) problem. The MLST consists of finding a spanning tree on a connected graph with the maximum number of leaves. We consider the leaves as customers and the internal vertices as AFSs and the depot.

This paper is organized as follows. Section 2 gives a literature review concerning the available G-VRP instances sets. Section 3 provides a formal definition of the problem. Section 4 describes the instance generation procedure. Section 5 presents the computational experiments and the generated instances sets. And finally, Sect. 6 presents the final remarks and possible future works.

2 Literature Review

Plenty of works have approached the G-VRP [1–3, 5, 9, 10, 15, 18, 20], as well as its variants for example, the Multiple Technologies and Partial Recharges (GVRP-

MTPR) [16], the Recharging VRP (RVRP) [11, 25], the Electric Capacitated Covering Tour Problem (ECCTP) [6], and many others [7, 13, 21].

The first G-VRP instances set was proposed by [15] and comprises 52 instances with 20 to 500 customers and 3 to 28 AFSs. Each instance was randomly generated assuming a grid of 330 by 300 miles. The majority of the instances in this set contain 20 customers and have been randomly generated to simulate different scenarios in terms of customer and AFSs distribution. The depot location was fixed and assumed to be located near the center of the grid in all scenarios. The larger instances [15] were created by using a medical textile supply company depot location in Virginia and a customer pool based on hospital locations in Virginia, and the District of Columbia. The refueling stations correspond to actual biodiesel stations located in the region. The number of customers in these larger instances ranges from 111 to 500. It should be mentioned that these instances may contain some infeasible customers. According to [15], a customer is infeasible if it cannot be visited by a route making at most one refueling stop. Motivated by the fact that most of these instances are small, and the five larger instances with 111 customers only differ in the number of refueling stations (ranging from 21 to 28), [1] proposed a new set of G-VRP instances. This set contains four additional instances by extracting the first 75 and 100 customers from the two 111-customer instances having 21 and 28 stations. Moreover, [1] created another set of instances by extracting customers from the larger instances of [15]. The set AB [1] contains 40 instances with several customers ranging from 50 to 100, split into two subsets AB1 and AB2. To solve the AB1 instances, [1] assumed that all of the infeasible customers are removed a priori. Concerning the AB2 instances, they have the same customers and refueling stations as those in AB1, that is, AB2 instances contain customers that cannot be served with a single refueling stop. Reference [1] solve the AB2 instances by an exact approach that deals with consecutive refueling stops.

3 Problem Definition

In the G-VRP proposed by [15], a set of customers with known services times must be attended by a set of homogeneous alternative fuel vehicles, which operate from a common depot. These vehicles have a lower autonomy in terms of traveled distance, forcing them to stop and refuel in AFSs. Each customer must be visited once by exactly one vehicle and the time spent by a vehicle in a route cannot exceed a maximum operation time. The objective is to reduce the total traveled distance.

Consider a complete directed graph $G(V, E)$, where $V = C \cup F \cup \{v_0\}$, where C is the set of customers, and F is the set of AFSs. The $s_i \geqslant 0$ is the customer v_i service time, and $s_f \geqslant 0$ is the AFS v_f refueling time. Vertex $v_0 \in V$ is the depot and E is the set of edges. Edges $(i, j) \in E$ have metric costs $c_{ij} > 0$. The $\rho > 0$ is the vehicle energy consumption rate, and $e_{ij} = c_{ij}\rho$ is the required energy to traverse edge (i, j). The $\xi > 0$ is the vehicle average speed, and $t_{ij} = \frac{c_{ij}}{\xi}$ is the required time to traverse the edge (i, j). In the depot, there is a set $M = \{1, \ldots, |C|\}$

of homogeneous vehicles with an autonomy of β and maximum operation time T. The goal of G-VRP consists in defining at most $|M|$ routes with minimum cost satisfying the following constraints:

- Each route begins and ends at the vertex depot v_0;
- Each vertex $v_i \in C$ is visited once by exactly one route;
- Each vertex $v_f \in F$ can be visited multiple times by multiple routes, thus, each route might have subcycles;
- The autonomy of a vehicle decreases in e_{ij} when it traverses edge $(i, j) \in E$. It is not possible for a vehicle to traverse an edge with a required energy higher than its autonomy. When the vehicle visits an AFS, its autonomy is restored to β;
- The time spent by a route must not exceed T.

4 Instance Generation Framework

We consider two types of instances, the first (I_1) not allowing consecutive AFS visits to serve a customer (as considered by [15]), and the second (I_2) allowing such property (as considered by the [1]'s instances set AB2). These instances are generated as solutions to a variant of the MLST.

Figure 1 contains three subfigures, Fig. 1a shows an example of an MLST instance, and the two remaining subfigures present examples of valid solutions for that instance, where the nodes in grey are the leaves. Figure 1b shows a non-optimal solution, whereas Fig. 1c shows the optimal solution.

In our context, an MLST instance is given by a graph with a distinct vertex (root) representing the depot. In the MLST solution, the leaves correspond to the

(a) (b) (c)

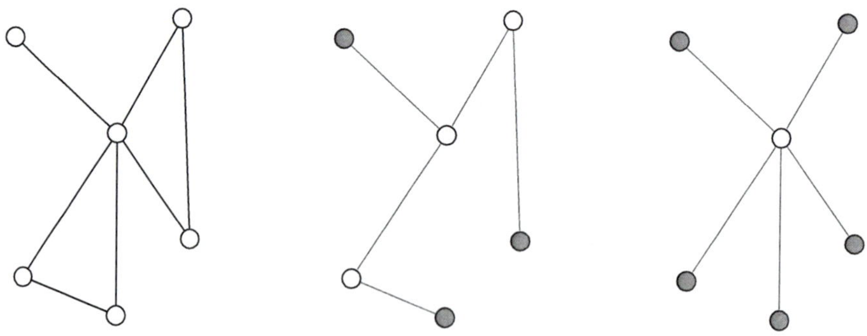

Fig. 1 An MLST instance and two examples of solutions. (**a**) An MLST instance with six nodes and seven edges. (**b**) An MLST solution with three leaves. (**c**) The MLST optimal solution with five leaves

G-VRP customers and the remaining nodes are the AFSs and depot. There are three main reasons to use the MLST problem as the framework to generate the set of instances. (i) the generated instances ensure that every customer can be reached by a vehicle starting at the depot and visiting one or more AFSs, (ii) the generated instances maximize the number of customers that can be served while maintaining feasibility, and (iii) despite the MLST being NP-hard, the literature provides fast exact approaches based on ILP.

In the flow-based model proposed by [22] the MLST is reduced to the Maximum Leaf Spanning Arborescence (MLSA) problem by replacing each edge (u, v) from the original graph $G(V, E)$ by two oppositely directed arcs (u, v) and (v, u) obtaining a digraph $D = (V, A)$. In this case, V contains the customers with the depot, and A the set of oppositely directed arcs between all the nodes in V. The model considers a new vertex v_t (terminal) along with a set of artificial arcs $\{(i, t) : v_i \in V \backslash \{v_r\}\}$ where v_r is the root, which in our case is v_0 (the depot). Let $D^{'} = (V^{'}, A^{'})$ be the resulting digraph, and let $c_{ij} > 0 : (i, j) \in E$ be the cost of edge $(i, j) \in E$.

Below is presented the MILP formulation F_{MLSA} used to generate the set of instances I_1. The decision variable f_{ij} represents a flow in arc $(i, j) \in A^{'} : v_i \neq v_t \wedge v_j \neq v_t$, and $f_{it} = 1$ if vertex $v_i \in V \backslash \{v_0\}$ is a leaf in the solution. The user-defined parameter r represents the number of AFSs (internal vertices) of the solution.

$$\text{Max } s = \sum_{v_i \in V \backslash \{v_0\}} f_{it} \tag{1}$$

s.t.

$$s \leqslant |V| - 1 - r \tag{2}$$

$$\sum_{v_j \in V} f_{0j} = |V| - 1 + s \tag{3}$$

$$\sum_{v_j \in V} f_{ji} - \sum_{v_j \in V \backslash \{v_0\}} f_{ij} = 1 \qquad v_i \in V \backslash \{v_0\} \tag{4}$$

$$\sum_{v_j \in V \backslash \{v_0\}} f_{ij} + 2(|V| - 2) f_{it} \leqslant 2(|V| - 2) \qquad v_i \in V \backslash \{v_0\} \tag{5}$$

$$f_{ij} = 0 \qquad \forall v_i, v_j \in V : c_{ij} > \beta \tag{6}$$

$$\sum_{v_j \in V : c_{ij} \leqslant \frac{\beta}{2} \wedge c_{j0} \leqslant \beta} 1 - f_{jt} \geqslant f_{it} \qquad \forall v_i \in V \backslash \{v_0\} : c_{i0} > \frac{\beta}{2} \tag{7}$$

$$0 \leqslant f_{ij} \leqslant 2|V| - 2 \qquad v_i, v_j \in V \tag{8}$$

$$f_{it} \in \mathbb{B} \qquad (i, t) \in A^{'} \tag{9}$$

Constraint (2) forces the number of leaves (customers) to be less than or equal to $|V| - 1 - r$. Constraint (3) ensures that the root outgoing flow is equal to $|V| - 1$ (to attend all the demand of $V \backslash \{v_0\}$) plus the number of leaves s, since each leaf sends one unit of flow to the terminal. Constraint (4) says that each vertex distinct from v_0 and v_t consumes one unit of flow. Constraint (5) implies that if v_i is a leaf, then it can only send flow to the terminal. Constraint (6) forbids flow in arcs with cost greater then β. Constraint (7) prohibits any leaf (customer) to be served after two or more consecutive AFS visits. Constraints (8) state that flows are positive and bounded by $2|V| - 2$. Constraints (9) define the domain constraints for $f_{it} : (i, t) \in A'$. The set of instance I_2 was generated by replacing the constraints (7) by the constraints (10), that guarantees for every customer that there will be at least one AFS in the range of half the autonomy of a fully charged vehicle:

$$\sum_{v_j \in V : c_{ij} \leqslant \frac{\beta}{2}} 1 - f_{jt} \geqslant f_{it} \qquad \forall v_i \in V \backslash \{v_0\} : c_{i0} > \frac{\beta}{2} \tag{10}$$

Suitable choices were made for the values of the parameters, as shown below. It is worth pointing out that the instance generation framework allows adjusting these parameters to conform with the underlying application. The vehicle average speed was defined as being the value necessary to traverse in one unit of time the instance longest edge:

$$\xi = max_{v_i, v_j \in V} c_{ij} \tag{11}$$

The AFS service time was defined as equal to one unit of time:

$$s_f = \frac{max_{v_i, v_j \in V} c_{ij}}{\xi} = 1 \qquad \forall v_f \in F \tag{12}$$

That is, an AFSs service time is equals to the traversing time of the longest graph edge, since we want the refueling time to be relatively high. The customer's service time was defined as:

$$s_i = \frac{\text{random}_{v_i, v_j \in V} c_{ij}}{\xi} \qquad \forall v_i \in F \tag{13}$$

That is, a customer service time is equals to a random graph edge traversing time, since we want to include diversity in the customers service time. It is noteworthy that the services times of every node and traversing time of every edge are normalized in the interval (0, 1].

The route time limit T was defined as:

$$T = \lceil 2\alpha \rceil \tag{14}$$

Where $\alpha = max_{v_i \in C}\lambda_i$ represents the maximum λ_i for every $v_i \in C$, and λ_i represents the minimum time feasible route containing only AFSs, the depot, and the customer $v_i \in C$.

From each MLST instance, three G-VRP instances were generated, varying the number of AFSs in the floor of 10%, 20%, and 30% of the total number of vertices (excluding the depot). Algorithm 1 gives the pseudocode of the instance generation framework.

Algorithm 1: G-VRP instance generation framework

Data: An MLST instance I and an AFS rate r.
Result: A G-VRP instance $I^{'}$.

1 $i \leftarrow 0$;
2 $j \leftarrow \sum_{a \in A} c_a$;
3 Let $I^{'} \leftarrow nil$ be an empty G-VRP instance;
4 **while** $i \leqslant j$ **do**
5     $\beta \leftarrow \lfloor \frac{i+j}{2} \rfloor$;
6   **if** F_{MLSA} *with* β *and* r *is feasible* **then**
7   Let S be a F_{MLSA} with β and r;
8         $I^{'} \leftarrow$ a G-VRP solution with customers as the leaves of S, with depot as the root of S, with AFSs as the remaining nodes of S, with the vehicle fuel capacity as β, and following the equations (12–14);
9   Let $r^{'}$ be the number of AFSs given by the $I^{'}$;
10   **if** $r^{'} = r$ **then**
11             $j \leftarrow \beta - 1$;
12   **continue**;
13   **end**
14   **end**
15     $i \leftarrow \beta + 1$;
16 **end**

The above algorithm does a binary search in $[0, \sum_{a \in A} c_a]$ to find the smallest integer value for β that produces a $MLSA$ feasible solution. The steps (1–2) declare and initialize the binary search leftmost and rightmost pointers i and j. In step (4) we have the binary search base case condition. Step (5) declares and initializes β as the ceil of the value at the middle of the rightmost and leftmost pointers. In the steps (11) and (15) the pointer j is updated to $\beta - 1$ if $r^{'} = r$, and the leftmost pointer i is updated to $\beta + 1$ if there is no feasible solution or if $r^{'} > r$.

The proposed framework is very customizable to other configurations, allowing straightforward changes to the graph structure (by changing the input graph), the AFSs rate r, the β binary search bounds (steps (1–2) of Algorithm 1), the vertices service times (12–13), and the vehicle time limit T (14).

5 Computational Experiments

The set P of VRP instances [8] were adopted as the instances for the MLST. These instances define their edges weight as $c_{ij} = \lfloor x + \frac{1}{2} \rfloor$, i.e., the closest integer to x, for each edge (i, j), where x is the Euclidean distance between vertices v_i and v_j.

The machine used to execute the experiments was an Intel(R) Xeon(R) CPU E5-2420 @1.90GHz, with 32 GB of RAM, and with 1TB of HD. The operational system used was the Ubuntu 18.04.4 LTS (bionic) 64 bits. The programming language used to generate the instances and to run the proposed models was C++. The source code is available at [4]. The MILP solver used was CPLEX 12.10 with ILOG Concert Technology using the Academic license.

Table 1 shows the G-VRP generated instances. As we can see the proposed framework is very fast, resulting in execution times of at most six seconds for every instance. An interesting observation is that in Augerat's VRP instances [8] with the same number of customers, the trade-offs among parameters β, $|F|$ and T become apparent. In particular, as more AFSs are allocated, the smaller β becomes, forcing the vehicles to make more stops to refuel. These trade-offs analysis could assist in the decision making process of AFS allocation of a city, for example, given the estimates of the average vehicles autonomies.

Table 1 Sets I_1 and I_2 of instances

	I_1					I_2				
[8] 's instance	$\|C\|$	$\|F\|$	β	T	CPU(s)	$\|C\|$	$\|F\|$	β	T	CPU(s)
P-n16-k8	13	2	55	6	0,23	13	2	55	6	0,22
	11	4	25	10	0,16	12	3	30	7	0,17
	10	5	23	10	0,25	10	5	25	7	0,21
P-n19-k2	17	1	60	8	0,27	17	1	60	8	0,22
	14	4	25	13	0,28	15	3	32	8	0,29
	13	5	21	17	0,25	13	5	30	7	0,2
P-n20-k2	17	2	55	8	0,24	17	2	55	8	0,22
	15	4	25	13	0,22	16	3	32	8	0,28
	14	5	21	17	0,2	14	5	30	7	0,19
P-n21-k2	18	2	40	8	0,2	18	2	40	8	0,19
	15	5	23	12	0,24	16	4	32	8	0,29
	13	7	19	14	0,46	13	7	25	7	0,19
P-n22-k2	19	2	40	8	0,18	19	2	40	8	0,23
	16	5	23	12	0,3	17	4	32	8	0,31
	14	7	19	22	0,23	14	7	25	7	0,18
P-n23-k8	20	2	40	8	0,3	20	2	40	8	0,35
	17	5	23	12	0,23	18	4	32	8	0,23
	15	7	19	16	0,32	15	7	25	7	0,24

(continued)

Table 1 (continued)

	I_1					I_2				
[8] 's instance	$\|C\|$	$\|F\|$	β	T	CPU(s)	$\|C\|$	$\|F\|$	β	T	CPU(s)
P-n40-k5	35	4	51	7	0,63	35	4	51	7	0,62
	31	8	27	11	0,68	32	7	32	7	0,64
	27	12	23	11	0,56	27	12	27	6	0,58
P-n45-k5	40	4	42	7	0,73	40	4	42	7	0,8
	35	9	25	11	0,78	35	9	30	6	0,73
	29	15	21	20	1,15	30	14	27	6	1,04
P-n50-k7	45	4	42	7	0,86	45	4	42	7	0,96
	39	10	25	11	0,9	39	10	27	7	0,86
	32	17	19	15	0,9	33	16	25	7	0,93
P-n51-k10	45	5	36	7	0,9	45	5	36	7	0,91
	37	13	23	11	0,92	40	10	30	6	0,97
	31	19	19	19	0,96	33	17	25	6	0,84
P-n55-k7	49	5	32	7	1,05	49	5	32	7	1,13
	43	11	23	16	1,13	40	14	25	7	0,99
	33	21	17	15	0,96	38	16	25	7	1,09
P-n60-k10	54	5	36	7	1,67	54	5	36	7	2,06
	47	12	23	13	1,49	47	12	27	6	1,38
	42	17	19	15	1,55	42	17	25	6	1,36
P-n65-k10	58	6	36	7	1,87	58	6	36	7	1,94
	50	14	23	16	1,63	50	14	27	6	1,38
	40	24	17	15	1,96	45	19	25	6	1,77
P-n70-k10	63	6	36	7	2,28	63	6	36	7	2,24
	54	15	23	25	2,01	54	15	27	6	1,8
	48	21	19	20	2,06	48	21	25	6	1,87
P-n76-k4	68	7	32	12	2,6	68	7	32	7	2,42
	58	17	21	13	2,8	59	16	27	6	1,99
	51	24	17	19	2,72	52	23	25	6	2,41
P-n101-k4	88	12	25	11	5,23	90	10	32	7	5,06
	78	22	19	15	5,35	80	20	25	6	4,42
	65	35	15	19	4,97	68	32	23	6	4,17

6 Conclusions

The presented work addresses the lack of G-VRP instances in the literature by proposing an effective framework to generate feasible G-VRP instances. The proposed method is highly customizable, allowing the generated instances to conform with the underlying application and to easily change the minimum number of AFSs, the vehicle properties (11), (14), and the vertices service times (12–13).

As future research, we consider generalizing the proposed framework to other variants of VRP, by including other constraints such as time windows, vehicle capacity, multiple depots, and others. An extended framework should also be considered to recommend alternative graph structures for the instance.

References

1. Andelmin, J., Bartolini, E.: An exact algorithm for the green vehicle routing problem. Transportation Science (2017). https://doi.org/10.1287/trsc.2016.0734
2. Andelmin, J., Bartolini, E.: A multi-start local search heuristic for the Green Vehicle Routing Problem based on a multigraph reformulation. Comput. Oper. Res. (2019). https://doi.org/10.1016/j.cor.2019.04.018
3. Andrade, M.D.: Formulations for the green vehicle routing problem. Institute of Computing, University of Campinas, Campinas, São Paulo, Brazil (2020)
4. Andrade, M.D.: Framework for GVRP instance generation. In: GitHub (2020). https://github.com/My-phd-degree/G-VRP-instance-generation. Cited 01 Apr 2021
5. Andrade, M.D., Usberti, F.L.: Valid Inequalities for the Green Vehicle Routing Problem. Anais do V Encontro de Teoria da Computação (2020). https://doi.org/10.5753/etc.2020.11086
6. Arakaki, R.K., Maziero, L.P., Andrade, M.D., Hama, V.M.F., Usberti, F.L.: Routing electric vehicles with remote servicing. Model. Optim. Green Logist. (2020). https://doi.org/10.1007/978-3-030-45308-4_8
7. Asghari, M., Mirzapour Al-e-hashem, S. M. J.: Green vehicle routing problem: A state-of-the-art review. Int. J. Prod. Econ. (2021). https://doi.org/10.1016/j.ijpe.2020.107899
8. Augerat, P.: Polyhedral approach of the vehicle routing problem. Institut National Polytechnique de Grenoble - INPG (1995). https://tel.archives-ouvertes.fr/tel-00005026. Cited 01 Apr 2021
9. Bo, P., Yuan, Z., Yuvraj, G., Xiding, C.: A memetic algorithm for the green vehicle routing problem. Sustainability (2019). https://doi.org/10.3390/su11216055
10. Bruglieri, M., Mancini, S., Pezzella, F., Pisacane, O.: A path-based solution approach for the green vehicle routing problem. Comput. Oper. Res. (2019). https://doi.org/10.1016/j.cor.2018.10.019
11. Conrad, R.G., Figliozzi, M.A.: The recharging vehicle routing problem. In: Proceedings of the 2011 Industrial Engineering Research Conference (2011). https://doi.org/10.1016/j.cor.2016.03.013
12. Ćirović, G., Pamuçar, D., Božanić, D.: Green logistic vehicle routing problem: Routing light delivery vehicles in urban areas using a neuro-fuzzy model. Expert Syst. Appl. (2014). https://doi.org/10.1016/j.eswa.2014.01.005
13. Das, K., Das, R.: Green vehicle routing problem: A critical survey. Intell. Tech. Appl. Sci. Technol., 736–745 (2020)
14. Dod, J.: Sources of greenhouse gas emissions. In: The Dictionary of Substances and Their Effects. United States Environmental Protection Agency (2020). https://www.epa.gov/ghgemissions/sources-greenhouse-gas-emissions. Cited 01 Apr 2021
15. Erdoğan, S., Miller-Hooks, E.: A green vehicle routing problem. Transport. Res. E Logist. Transport. Rev. (2012). https://doi.org/10.1016/j.tre.2011.08.001
16. Felipe, A., Ortuño, M.T., Righini, G., Tirado, G.: A heuristic approach for the green vehicle routing problem with multiple technologies and partial recharges. Transport. Res. E Logist. Transport. Rev. (2014). https://doi.org/10.1016/j.tre.2014.09.003
17. Jun, Y., Hao, S.: Battery swap station location-routing problem with capacitated electric vehicles. Comput. Oper. Res. (2015). https://doi.org/10.1016/j.cor.2014.07.003
18. Koç, Ç., Karaoglan, I.: The green vehicle routing problem: A heuristic based exact solution approach. Appl. Soft Comput. (2016). https://doi.org/10.1016/j.asoc.2015.10.064

19. Kuby, M., Lim, S.: Location of alternative-fuel stations using the flow-refueling location model and dispersion of candidate sites on arcs. Netw. Spat. Econ. (2007). https://doi.org/10.1007/s11067-006-9003-6
20. Leggieri, V., Haouari, M.: A practical solution approach for the green vehicle routing problem. Transport. Res. E Logist. Transport. Rev. (2017). https://doi.org/10.1016/j.tre.2017.06.003
21. Lin, C., Choy, K.L., Ho, G.T.S., Chung, S.H., Lam, H.Y.: Survey of green vehicle routing problem: Past and future trends. Expert Syst. Appl. (2014). https://doi.org/10.1016/j.eswa.2013.07.107
22. Reis, M.F., Lee, O., Usberti, F.L.: Flow-based formulation for the maximum leaf spanning tree problem. Electron. Notes Discrete Math. (2015). https://doi.org/10.1016/j.endm.2015.07.035
23. Ritchie, H., Roser, M.: CO_2 and other greenhouse gas emissions. In: The Dictionary of Substances and Their Effects. United States Environmental Protection Agency (2016). https://ourworldindata.org/co2-and-other-greenhouse-gas-emissions Cited 01 Apr 2021
24. Toth, P., Vigo, D.: Vehicle routing: problems, methods, and applications (2014)
25. Wang, Y.-W., Lin, C.-C., Lee, T.-J.: Electric vehicle tour planning. Transport. Res. D Transport Environ. (2018). https://doi.org/10.1016/j.trd.2018.04.016
26. Yeh, S.: An empirical analysis on the adoption of alternative fuel vehicles: The case of natural gas vehicles. Energy Policy (2007). https://doi.org/10.1016/j.enpol.2007.06.012

An Optimization Model for Service Requests Management in a 5G Network Architecture

Gabriella Colajanni and Daniele Sciacca

Abstract In this paper, we present a three-tier supply chain network model consisting of a fleet of UAVs organized as a FANET (Fly ad hoc network) connected one to each other with direct wireless links, managed by a fleet of UAV controllers, whose purpose is to provide 5G network slices on demand to users and devices on the ground. The aim of this paper is to determine the optimal distributions of request flows. We obtain a constrained optimization problem for which we derive the associated Variational inequality formulation. Also, qualitative properties in terms of existence and uniqueness of solution are provided. Finally, a numerical example is performed to validate the effectiveness of the model.

Keywords Variational inequality · Service requests management · UAVs · Three-tier supply chain

1 Introduction

In the last years the impact of 5G technologies is revolutionizing all social and economic sectors. Therefore, several studies have focused on mobile networks and the basic idea is to create virtual logical networks, called "Network Slices", which share the same physical access and transport infrastructure, to support different application cases with particular characteristics and requirements. The same network can have multiple slices, each dedicated to specific services or customers. This infrastructure guarantees greater flexibility, efficient use of resources and greater business opportunities. In this paper we consider some users and devices on the ground which require different application or network services (see [9] for an example based on Video Monitoring). Such service requests are managed and executed by a fleet of Unmanned Aerial Vehicles (UAVs) organized as a Flying Ad

G. Colajanni (✉) · D. Sciacca
Department of Mathematics and Computer Science, University of Catania, Catania, Italy
e-mail: colajanni@dmi.unict.it; daniele.sciacca@unipa.it

A. Masone et al. (eds.), *Optimization and Data Science: Trends and Applications*, AIRO Springer Series 6, https://doi.org/10.1007/978-3-030-86286-2_7

hoc Network (FANET) whose purpose is to provide 5G network slices (see [10]), reaching also remote rural areas (see [3] and [1] for a UAV-Based Cellular Networks managing problem in rural areas and in which authors propose a spatial and a temporal decomposition). Particularly, we suppose that the UAVs are distributed in two different layers: the controller layer and the execution layer. The controller UAVs acquire service requests from users and devices on the ground and send them to the UAVs of the fleet on the upper level, where the services are performed.

In this paper we suppose also the possibility of the provider to add additional drones to perform the services requested on demand and we also take into account the quality of the network in providing services.

So, the purpose of this paper is to propose a constrained optimization model, aimed at maximizing the provider's profit, given by the total revenue obtained from the sale of services to users and devices on the ground to which all transmission and execution costs are subtracted, as well as the costs for additional UAVs are subtracted while the quality in terms of profit is summed (see [4] for an optimization model of IaaS provider). Of course, the model proposed in this paper takes into account the managing and execution capabilities of each UAV, the conservation laws related to the supply chain network and, a budget and a Quality of Service (QoS) constraints (see [8]).

Therefore, the proposed optimization model allows us to determine the equilibrium flows of the quantity of services requested by each user or device to each controller UAV, the quantity of service requests that each controller UAV sent to each UAV of the fleet, the quantity of executed services by each UAV, if and which additional UAVs to activate.

The rest of this paper is organized as follows. In Sect. 2 we present the mathematical model and derive the optimality conditions of the provider, who desires to maximize its profit. In Sect. 3 we provide a variational formulation of the problem and qualitative properties in terms of existence and uniqueness of the solution. To validate the effectiveness of the model, in Sect. 4 we perform an illustrative numerical example. Section 5 is dedicated to the conclusions.

2 The Model

The supply chain network, consisting of a fleet of UAVs, controller UAVs and users or devices on the ground, is depicted in Fig. 1. The typical user or device on the ground is denoted by g, $g = 1, \dots, G$, and could require K types of services (network services or application services). The requests for service k, $k = 1, \dots, K$, from users and devices are received by the controller UAVs which are spatially distributed in the considered geographical area. Each controller UAV u, $u = 1, \dots, U$, sends the requests for execution of the services to the fleet of UAVs at the higher level.

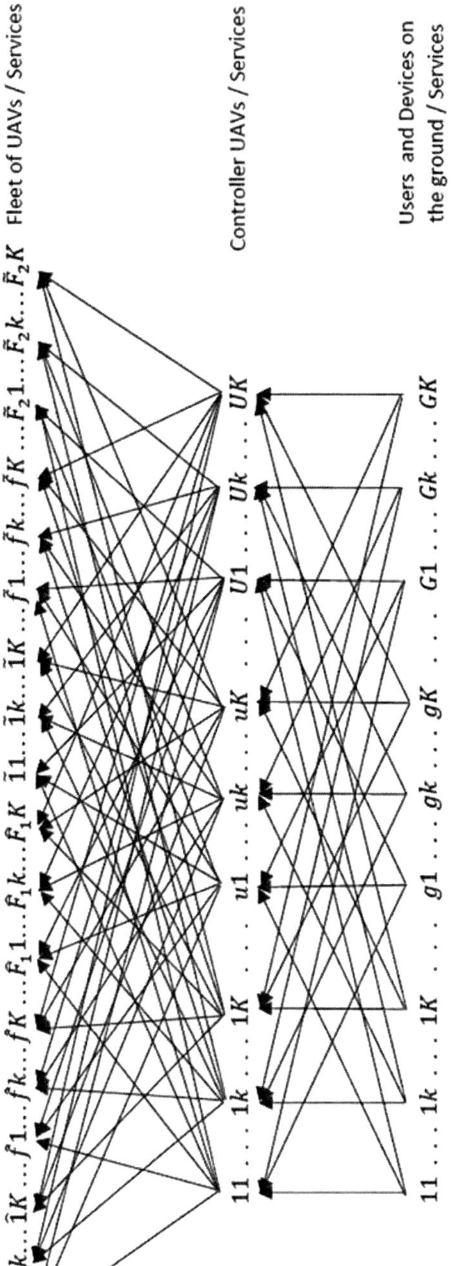

Fig. 1 Network topology

The fleet of UAVs consists of $\hat{F}_1$ pre-existing UAVs (the typical one is denoted by $\hat{f}$), to which the provider can join $\tilde{F}_2$ additional UAVs. Therefore, we consider the following sets:

- $\hat{\mathcal{F}}_1 = \{\hat{1}, \ldots, \hat{f}, \ldots, \hat{F}_1\}$, the set of pre-existing UAVs;
- $\tilde{\mathcal{F}}_2 = \{\tilde{1}, \ldots, \tilde{f}, \ldots, \tilde{F}_2\}$, the set of possible additional UAVs.

Each UAV of the upper level receives the service requests from the controller UAVs and performs the executions.

In this paper we analyze the supply chain network using a system-optimization approach in order to find the global optimization problem. Therefore, the purpose of this paper is to determine a model that aims to maximize the total profit (defined by the difference between the revenue and the sum of costs) and the quality.

The variables and the parameters of the model are described in Tables 1 and 2, respectively.

Now, we present the cost and quality functions. Let:

- c_{gu} be the transmission cost of the service requests from user or device g to controller UAV u and let us assume c_{gu} is a function of $\sum_{k=1}^{K} x_{guk}$:

$$c_{gu} = c_{gu}\left(\sum_{k=1}^{K} x_{guk}\right) = c_{gu}(X_{gu}), \quad \forall g = 1, \ldots, G, \ \forall u = 1, \ldots, U.$$

In particular, for these functions we assume the following generic quadratic expression:

$$c_{gu}(X_{gu}) = \eta_{gu}\left(\sum_{k=1}^{K} x_{guk}\right)^2 + \eta'_{gu}\sum_{k=1}^{K} x_{guk}, \quad \forall g, \ \forall u.$$

Table 1 Variables for the model

Notation	Variables
$x_{guk} \geq 0$	The quantity of service k requested by user or device g on the ground to the controller UAV u and let us group all these quantities into the vectors X_{gu} for all k and X for all g and for all u
$y_{u\hat{f}k} \geq 0$	The quantity of service k requests sent by the controller UAV u to the pre-existing UAV $\hat{f} \in \hat{\mathcal{F}}_1$ belonging to the upper tier fleet and let us group all these quantities into the vectors: $Y_{u\hat{f}}$ for all k, $Y_{\hat{f}k}$ for all u, $Y_{\hat{f}}$ for all u and for all k and Y^2 for all u, $\hat{f}$ and k
$z_{u\tilde{f}k} \geq 0$	The quantity of service k requests sent by the controller UAV u to the additional UAV $\tilde{f} \in \tilde{\mathcal{F}}_2$ which the provider can decide to activate and let us group all these quantities into the vectors: $Z_{u\tilde{f}}$ for all k, $Z_{\tilde{f}}$ for all u and k and Z for all u, $\tilde{f}$ and k

Table 2 Parameters for the model

Notation	Parameters
R_{gk}	The demand for service (application service or network service) k from the user or device g on the ground, $\forall g = 1, \ldots, G,\ k = 1, \ldots, K$
$\overline{S}_u$	The maximum capacity related to the controller UAV u, that is the maximum number of service requests that the controller UAV u is able to manage, $\forall u = 1, \ldots, U$
s_k	The execution space requested to perform service k, $\forall k = 1, \ldots, K$
S_f	The maximum capacity related to the UAV f, $\forall f \in \mathcal{F}_3$, that is the maximum execution space that the (pre-existing or additional) UAV f can bear
$\overline{B}$	The maximum available budget for additional UAVs at the highest level of the network
α_k	The unit benefit associated with the quality of service k, $\forall k = 1, \ldots, K$, and it can be intended as a parameter that allows us to express the quality in terms of profit (same unit of measure of the cost functions); hence, the gain that the provider receives and which it aims to maximize is given by the product between α_k and the quality: $$\alpha_k q_k \left(\sum_{u=1}^{U} \sum_{\hat{f} \in \hat{\mathcal{F}}_1} y_{u\hat{f}k} + \sum_{u=1}^{U} \sum_{\tilde{f} \in \tilde{\mathcal{F}}_2} z_{u\tilde{f}k} \right);$$ note that the parameter α_k enables distinct services to have different profits associated to their quality, based on their importance and needs
Q_k	The minimum quality of the service (QoS) k, $\forall k = 1, \ldots, K$, previously established in a service-level agreement (SLA), that the provider has to guarantee
ρ_k	The revenue obtained for a unit of service k executed, $\forall k = 1, \ldots, K$

- $c_{u\hat{f}}$ and $c_{u\tilde{f}}$ be the transmission cost of the service requests from controller UAV u to any pre-existing UAV $\hat{f} \in \hat{\mathcal{F}}_1$ and any additional UAV $\tilde{f} \in \tilde{\mathcal{F}}_2$, respectively, and let us assume $c_{u\hat{f}}$ is a function of $\sum_{k=1}^{K} y_{u\hat{f}k}$ and $c_{u\tilde{f}}$ is a function of $\sum_{k=1}^{K} z_{u\tilde{f}k}$:

$$c_{u\hat{f}} = c_{u\hat{f}} \left(\sum_{k=1}^{K} y_{u\hat{f}k} \right), \quad c_{u\tilde{f}} = c_{u\tilde{f}} \left(\sum_{k=1}^{K} z_{u\tilde{f}k} \right) \quad \forall u,\ \forall \hat{f} \in \hat{\mathcal{F}}_1,\ \forall \tilde{f} \in \tilde{\mathcal{F}}_2$$

that is $c_{u\hat{f}} = c_{u\hat{f}}(Y_{u\hat{f}}),\ \forall u = 1, \ldots, U,\ \forall \hat{f} \in \hat{\mathcal{F}}_1$ and $c_{u\tilde{f}} = c_{u\tilde{f}}(Z_{u\tilde{f}}),\ \forall u = 1, \ldots, U,\ \forall \tilde{f} \in \tilde{\mathcal{F}}_2$. For the aforementioned functions, the following general quadratic expression is assumed:

$$c_{u\hat{f}}(Y_{u\hat{f}}) = \kappa_{u\hat{f}} \left(\sum_{k=1}^{K} y_{u\hat{f}k} \right)^2 + \kappa'_{u\hat{f}} \sum_{k=1}^{K} y_{u\hat{f}k}, \quad \forall u,\ \hat{f} \in \hat{\mathcal{F}}_1,$$

$$c_{u\tilde{f}}(Z_{u\tilde{f}}) = \mu_{u\tilde{f}} \left(\sum_{k=1}^{K} z_{u\tilde{f}k} \right)^2 + \mu'_{u\tilde{f}} \sum_{k=1}^{K} z_{u\tilde{f}k}, \quad \forall u,\ \tilde{f} \in \tilde{\mathcal{F}}_2,$$

- $c^{(E)}_{\hat{f}}$ and $c^{(E)}_{\tilde{f}}$ be the execution costs of requested services to the pre-existing UAV $\hat{f} \in \hat{\mathcal{F}}_1$ and to the additional UAV $\tilde{f} \in \tilde{\mathcal{F}}_2$, respectively, and let us assume $c^{(E)}_{\hat{f}}$ is a function of the total amount of executed services by pre-existing UAVs, $\sum_{u=1}^{U} \sum_{k=1}^{K} y_{u\hat{f}k}$ and $c^{(E)}_{\tilde{f}}$ is a function of the total amount of executed services by additional UAVs, $\sum_{u=1}^{U} \sum_{k=1}^{K} z_{u\tilde{f}k}$:

$$c^{(E)}_{\hat{f}} = c^{(E)}_{\hat{f}} \left(\sum_{u=1}^{U} \sum_{k=1}^{K} y_{u\hat{f}k} \right) = c^{(E)}_{\hat{f}} \left(Y_{\hat{f}} \right), \quad \forall \hat{f} \in \hat{\mathcal{F}}_1,$$

$$c^{(E)}_{\tilde{f}} = c^{(E)}_{\tilde{f}} \left(\sum_{u=1}^{U} \sum_{k=1}^{K} z_{u\tilde{f}k} \right) = c^{(E)}_{\tilde{f}} \left(Z_{\tilde{f}} \right), \quad \forall \tilde{f} \in \tilde{\mathcal{F}}_2.$$

In particular, we assume that such functions have the following quadratic expression:

$$c^{(E)}_{\hat{f}} \left(Y_{\hat{f}} \right) = \delta \left(\sum_{u=1}^{U} \sum_{k=1}^{K} y_{u\hat{f}k} \right)^2 + \delta' \sum_{u=1}^{U} \sum_{k=1}^{K} y_{u\hat{f}k},$$

$$c^{(E)}_{\tilde{f}} \left(Z_{\tilde{f}} \right) = \epsilon \left(\sum_{u=1}^{U} \sum_{k=1}^{K} z_{u\tilde{f}k} \right)^2 + \epsilon' \sum_{u=1}^{U} \sum_{k=1}^{K} z_{u\tilde{f}k};$$

- $c_{\tilde{f}}$ be the cost due to add a new UAV $\tilde{f} \in \tilde{\mathcal{F}}_2$ at the highest level of the network and let us assume $c_{\tilde{f}}$ is a function of the flow of requests received:

$$c_{\tilde{f}}\left(\sum_{u=1}^{U}\sum_{k=1}^{K} z_{u\tilde{f}k}\right) = c_{\tilde{f}}\left(Z_{\tilde{f}}\right), \quad \forall \tilde{f} \in \tilde{\mathcal{F}}_2,$$

we will assume this function in such a way that it is null in the event that no service request is sent to the UAV $\tilde{f}$ ($c_{\tilde{f}}(0) = 0$). The generic quadratic expression for these functions is as follows:

$$c_{\tilde{f}}(Z_{\tilde{f}}) = \beta\left(\sum_{u=1}^{U}\sum_{k=1}^{K} z_{u\tilde{f}k}\right)^2 + \beta'\left(\sum_{u=1}^{U}\sum_{k=1}^{K} z_{u\tilde{f}k}\right), \quad \forall \tilde{f} \in \tilde{\mathcal{F}}_2,$$

- q_k be the quality function related to the service k, $\forall k = 1, \ldots, K$, and let us assume q_k is an increasing function of the executed services $\sum_{u=1}^{U}\sum_{\hat{f}\in\mathcal{F}_1} y_{u\hat{f}k} + \sum_{u=1}^{U}\sum_{\tilde{f}\in\mathcal{F}_2} z_{u\tilde{f}k}$:

$$q_k = q_k\left(\sum_{u=1}^{U}\sum_{\hat{f}\in\mathcal{F}_1} y_{u\hat{f}k} + \sum_{u=1}^{U}\sum_{\tilde{f}\in\mathcal{F}_2} z_{u\tilde{f}k}\right) = q_k\left(Y_{\hat{f}}, Z_{\tilde{f}}\right), \quad \forall k = 1, \ldots, K.$$

Particularly, we will assume such a function depends on the difference between the total demand for service k that all the users and devices require and the amount of executed service k that they receive from all UAVs belonging to the fleet, $\sum_{g=1}^{G} R_{gk} - \left(\sum_{u=1}^{U}\sum_{\hat{f}\in\mathcal{F}_1} y_{u\hat{f}k} + \sum_{u=1}^{U}\sum_{\tilde{f}\in\mathcal{F}_2} z_{u\tilde{f}k}\right)$. For the quality function, we assume the following general expression:

$$q_k\left(Y_{\hat{f}}, Z_{\tilde{f}}\right) = \sum_{g=1}^{G} R_{gk} - \frac{\left(\sum_{g=1}^{G} R_{gk} - \left(\sum_{u=1}^{U}\sum_{\hat{f}\in\mathcal{F}_1} y_{u\hat{f}k} + \sum_{u=1}^{U}\sum_{\tilde{f}\in\mathcal{F}_2} z_{u\tilde{f}k}\right)\right)^2}{\sum_{g=1}^{G} R_{gk}}, \quad \forall k.$$

We observe that if $\sum_{u=1}^{U}\sum_{\hat{f}\in\mathcal{F}_1} y_{u\hat{f}k}+\sum_{u=1}^{U}\sum_{\tilde{f}\in\mathcal{F}_2} z_{u\tilde{f}k}=\sum_{g=1}^{G} R_{gk}$, i.e. the total demand for service k that all users and devices require its equal to the amount of executed service k that they receive from all UAVs belonging to the fleet, then the quality reaches the maximum value $(\sum_{g=1}^{G} R_{gk})$. If, instead, $\sum_{u=1}^{U}\sum_{\hat{f}\in\mathcal{F}_1} y_{u\hat{f}k}+\sum_{u=1}^{U}\sum_{\tilde{f}\in\mathcal{F}_2} z_{u\tilde{f}k}=0$, i.e. the amount of executed service k that users and devices receive from all UAVs belonging to the fleet is null, but $\sum_{g=1}^{G} R_{gk}>0$, then the quality of services provider reaches the minimum value 0. Finally, if $\sum_{u=1}^{U}\sum_{\hat{f}\in\mathcal{F}_1} y_{u\hat{f}k}+\sum_{u=1}^{U}\sum_{\tilde{f}\in\mathcal{F}_2} z_{u\tilde{f}k}<\sum_{g=1}^{G} R_{gk}$, the function quality increases with the executed services $\sum_{u=1}^{U}\sum_{\hat{f}\in\mathcal{F}_1} y_{u\hat{f}k}+\sum_{u=1}^{U}\sum_{\tilde{f}\in\mathcal{F}_2} z_{u\tilde{f}k}$, as supposed before.

Note that in this paper we also take into account the limitation that distinguishes UAVs, that is the limited flight duration due to the consumption of batteries and that we integrated in the maximum capacity of each UAV and in the execution cost of requested services, indeed we are assuming that each execution has a cost in terms of battery power because this causes a reduction of the flight duration. No cost of handling requests, at the level of controller UAVs, is considered in this paper, since we consider them negligible and we assume the cost to keep the UAVs u in flight constant and therefore we do not include it in our model, but extension to a more general case is easy.

As previously mentioned, we provide a system-optimization perspective for the entire supply chain network, then, we analyze the system from the point of view of the network and service provider. Therefore, the presented model aims at determining the optimal distributions of services requests flows. The objective function consists of the profit to maximize and is given by the total revenue obtained from the sale of services to users and devices on the ground, to which all transmission and execution costs are subtracted, as well as the costs for additional UAVs, while the quality in terms of profit is summed.

The problem formulation is as follows:

$$
\begin{aligned}
max \sum_{u=1}^{U}\sum_{k=1}^{K} \rho_k \cdot \left(\sum_{\hat{f}\in\hat{\mathcal{F}}_1} y_{u\hat{f}k} + \sum_{\tilde{f}\in\tilde{\mathcal{F}}_2} z_{u\tilde{f}k} \right) - \sum_{g=1}^{G}\sum_{u=1}^{U} c_{gu} \left(\sum_{k=1}^{K} x_{guk} \right) - \sum_{u=1}^{U}\sum_{\hat{f}\in\hat{\mathcal{F}}_1} c_{u\hat{f}} \left(\sum_{k=1}^{K} y_{u\hat{f}k} \right) \\
- \sum_{u=1}^{U}\sum_{\tilde{f}\in\tilde{\mathcal{F}}_2} c_{u\tilde{f}} \left(\sum_{k=1}^{K} z_{u\tilde{f}k} \right) - \sum_{\hat{f}\in\hat{\mathcal{F}}_1} c_{\hat{f}}^{(E)} \left(\sum_{u=1}^{U}\sum_{k=1}^{K} y_{u\hat{f}k} \right) - \sum_{\tilde{f}\in\tilde{\mathcal{F}}_2} c_{\tilde{f}}^{(E)} \left(\sum_{u=1}^{U}\sum_{k=1}^{K} z_{u\tilde{f}k} \right) \\
- \sum_{\tilde{f}\in\tilde{\mathcal{F}}_2} c_{\tilde{f}} \left(\sum_{u=1}^{U}\sum_{k=1}^{K} z_{u\tilde{f}k} \right) + \sum_{k=1}^{K} \alpha_k q_k \left(\sum_{u=1}^{U}\sum_{\hat{f}\in\hat{\mathcal{F}}_1} y_{u\hat{f}k} + \sum_{u=1}^{U}\sum_{\tilde{f}\in\tilde{\mathcal{F}}_2} z_{u\tilde{f}k} \right)
\end{aligned} \tag{1}
$$

subject to

$$\sum_{u=1}^{U} x_{guk} = R_{gk} \quad \forall g = 1, \dots, G,\ \forall k = 1, \dots, K, \tag{2}$$

$$\sum_{k=1}^{K}\sum_{g=1}^{G} x_{guk} \leq \overline{S}_u \quad \forall u = 1, \dots, U, \tag{3}$$

$$\sum_{\hat{f}\in\hat{\mathcal{F}}_1} y_{u\hat{f}k} + \sum_{\tilde{f}\in\tilde{\mathcal{F}}_2} z_{u\tilde{f}k} \leq \sum_{g=1}^{G} x_{guk} \quad \forall u = 1, \dots, U,\ \forall k = 1, \dots, K, \tag{4}$$

$$\sum_{u=1}^{U}\sum_{k=1}^{K} s_k y_{u\hat{f}k} \leq S_{\hat{f}} \quad \forall \hat{f} \in \hat{\mathcal{F}}_1, \tag{5}$$

$$\sum_{u=1}^{U}\sum_{k=1}^{K} s_k z_{u\tilde{f}k} \leq S_{\tilde{f}} \quad \forall \tilde{f} \in \tilde{\mathcal{F}}_2, \tag{6}$$

$$\sum_{\tilde{f}\in\tilde{\mathcal{F}}_2} c_{\tilde{f}} \left(\sum_{u=1}^{U}\sum_{k=1}^{K} z_{u\tilde{f}k} \right) \leq \overline{B}, \tag{7}$$

$$q_k \left(\sum_{u=1}^{U}\sum_{\hat{f}\in\hat{\mathcal{F}}_1} y_{u\hat{f}k} + \sum_{u=1}^{U}\sum_{\tilde{f}\in\tilde{\mathcal{F}}_2} z_{u\tilde{f}k} \right) \geq Q_k \quad \forall k = 1, \dots, K, \tag{8}$$

$$
\begin{aligned}
& x_{guk}, y_{u\hat{f}k}, z_{u\tilde{f}k} \in \mathbb{R}_+ \\
& \forall g = 1, \dots, G,\ \forall u = 1, \dots, U,\ \forall \hat{f} \in \hat{\mathcal{F}}_1,\ \forall \tilde{f} \in \tilde{\mathcal{F}}_2,\ \forall k = 1, \dots, K.
\end{aligned} \tag{9}
$$

The first constraint, (2), represents the conservation law, according to which the quantity of service k requested by user or device g on the ground to all controller UAVs equals the demand R_{gk}.

Constraint (4) establishes that the quantity of service requests that the controller UAV u can receive is less than or equal to the maximum capacity. Observe that constraint (2) must be verified, therefore, the total capacity of controller UAVs are such as to satisfy the demand for services from all users and devices.

Constraint (4) states that the quantity of service k requests sent by the controller UAV u to all the pre-existing and additional UAVs is less than or equal to the quantity of service k requested by all users or devices on the ground to the controller UAV u.

The requests of services that each pre-existing and additional UAV $\hat{f}$ and $\tilde{f}$, $\forall \hat{f} \in \hat{\mathcal{F}}_1$ and $\forall \tilde{f} \in \tilde{\mathcal{F}}_2$, can receive must satisfy the capacity constraints (5) and (6) which establishes that the execution space must not exceed the maximum allowed.

Constraint (7) means that there is a budget limit, $\overline{B}$, which represents the maximum available budget for adding new UAVs at the highest level of the network.

Constraint (8) affirms that the QoS defined in the SLA is guaranteed.

The latest constraint family defines the domain of the variables of the problem.

3 Variational Formulation

We now provide a variational formulation of problem (1)–(9) (see [2, 5, 6] for variational formulations applied to other fields). The advantages of using this formulation are manifold. Firstly, we can use the well-known variational inequality theory which guarantees us results of existence and uniqueness of the solution, which we will provide below, and, secondly, through the use of the Lagrangian function we can relax the critical constraints (7) and (8) considering the associated Lagrange multipliers. Moreover, to the best of our knowledge, no one has ever studied the variational formulation of this 5G network problem before.

First, we make the following assumption:

Hp. Let all the involved cost functions be continuously differentiable and convex and the quality functions be continuously differentiable and concave. □

This assumption is justified by the concept of *diminishing marginal utility*, according to which averages are better than the extremes. Moreover, we note that the above assumptions on the cost functions and their expressions are not unreasonable since we are dealing with continuous variables. Similar assumptions have been made in several network-based optimization model (see, for instance, [5, 6, 12, 13]).

We have the following result (see, for instance, [13]).

Theorem 1 *A vector $(X, Y, Z) \in \mathbb{K}$ is an optimal solution to the problem (1)–(9) if and only if there exist the Lagrange multiplier vectors $\lambda^{1*} \in \mathbb{R}_+$ and $\lambda^{2*} \in \mathbb{R}_+^K$ such* *that the vector* $(X, Y, Z, \lambda^{1*}, \lambda^{2*}) \in \mathbb{K} \times \mathbb{R}_+ \times \mathbb{R}_+^K$ *is a solution to the variational inequality:*

Find $(X, Y, Z, \lambda^{1*}, \lambda^{2*}) \in \mathbb{K} \times \mathbb{R}_+ \times \mathbb{R}_+^K$ *such that:*

$$\sum_{g=1}^{G}\sum_{u=1}^{U}\sum_{k=1}^{K}\left[\frac{\partial c_{gu}(X^*_{gu})}{\partial x_{guk}}\right] \times (x_{guk} - x^*_{guk})$$
$$+\sum_{u=1}^{U}\sum_{\hat{f}\in\hat{\mathcal{F}}_1}\sum_{k=1}^{K}\Bigg[\frac{\partial c_{u\hat{f}}(Y_{u\hat{f}})}{\partial y_{u\hat{f}k}} + \frac{\partial c^{(E)}_{\hat{f}}\left(Y^*_{\hat{f}}\right)}{\partial y_{u\hat{f}k}} - \rho_k - \alpha_k \frac{\partial q_k(Y^*, Z^*)}{\partial y_{u\hat{f}k}}$$
$$-\lambda_k^{2*}\frac{\partial q_k(Y^*, Z^*)}{\partial y_{u\hat{f}k}}\Bigg] \times (y_{u\hat{f}k} - y^*_{u\hat{f}k})$$
$$+\sum_{u=1}^{U}\sum_{\tilde{f}\in\tilde{\mathcal{F}}_2}\sum_{k=1}^{K}\Bigg[\frac{\partial c_{u\tilde{f}}(Z^*_{u\tilde{f}})}{\partial z_{u\tilde{f}k}} + \frac{\partial c^{(E)}_{\tilde{f}}\left(Z^*_{\tilde{f}}\right)}{\partial z_{u\tilde{f}k}} + \frac{\partial c_{\tilde{f}}(Z^*_{\tilde{f}})}{\partial z_{u\tilde{f}k}} - \alpha_k \frac{\partial q_k(Y^*, Z^*)}{\partial z_{u\tilde{f}k}} - \rho_k$$
$$+\lambda^{1*}\frac{\partial c_{\tilde{f}}(Z^*_{\tilde{f}})}{\partial z_{u\tilde{f}k}} - \lambda_k^{2*}\frac{\partial q_k(Y^*, Z^*)}{\partial z_{u\tilde{f}k}}\Bigg] \times (z_{u\tilde{f}k} - z^*_{u\tilde{f}k})$$
$$+\left[\overline{B} - \sum_{\tilde{f}\in\tilde{\mathcal{F}}_2} c_{\tilde{f}}\left(Z^*_{\tilde{f}}\right)\right] \times (\lambda^1 - \lambda^{1*})$$
$$+\sum_{k=1}^{K}\left[q_k\left(Y^*, Z^*\right) - Q_k\right] \times (\lambda_k^2 - \lambda_k^{2*}) \geq 0,$$
$$\forall (X, Y, Z, \lambda^1, \lambda^2) \in \mathbb{K} \times \mathbb{R}_+ \times \mathbb{R}_+^K. \tag{10}$$

where

$$\mathbb{K} := \left\{(X, Y, Z) \in \mathbb{R}_+^{GUK+U\hat{F}_1K+U\tilde{F}_2K} : (2)-(6) \text{ hold true}\right\}. \tag{11}$$

We now put variational inequality (10) into standard form, that is: determine $\mathcal{X}^* \in \mathcal{K}$ such that:

$$\langle F(\mathcal{X}^*), \mathcal{X} - \mathcal{X}^*\rangle \geq 0, \quad \forall \mathcal{X} \in \mathcal{K} \tag{12}$$

where $\langle \cdot, \cdot \rangle$ denotes the inner product in the Euclidean space $\mathbb{R}_+^{GUK+U\hat{F}_1K+U\tilde{F}_2K+1+K}$, $\mathcal{X} \equiv (X^1, X^2, X^3, \lambda^1, \lambda^2)$, F is a given function from $\mathcal{K}$ to $\mathbb{R}_+^{GUK+U\hat{F}_1K+U\tilde{F}_2K+1+K}$ and $\mathcal{K}$ is a closed and convex set. We put $N = GUK + U\hat{F}_1K + U\tilde{F}_2K + 1 + K$ and we define the N-dimensional column vector $X = (X^1, X^2, X^3, \lambda^1, \lambda^2)$ and the N-dimensional column vector $F(\mathcal{X}) = (F^1(\mathcal{X}), F^2(\mathcal{X}), F^3(\mathcal{X}), F^4(\mathcal{X}), F^5(\mathcal{X}))$, where the (g, u, k)-th component, F^1_{guk}, of $F^1(\mathcal{X})$ is given by

$$F^1_{guk} = \frac{\partial c_{gu}(X_{gu})}{\partial x_{guk}},$$

the $(u, \hat{f}, k)$-th component, $F^2_{u\hat{f}k}$, of $F^2(\mathcal{X})$ is given by

$$F^2_{u\hat{f}k} = \frac{\partial c_{u\hat{f}}(Y_{u\hat{f}})}{\partial y_{u\hat{f}k}} + \frac{\partial c^{(E)}_{\hat{f}}\left(Y_{\hat{f}}\right)}{\partial y_{u\hat{f}k}} - \rho_k - \alpha_k \frac{\partial q_k(Y, Z)}{\partial y_{u\hat{f}k}} - \lambda_k^{2*} \frac{\partial q_k(Y, Z)}{\partial x_{u\hat{f}k}},$$

the $(u, \tilde{f}, k)$-th component, $F^3_{u\tilde{f}k}$, of $F^3(\mathcal{X})$ is given by

$$F^3_{u\tilde{f}k} = \frac{\partial c_{u\tilde{f}}(Z_{u\tilde{f}})}{\partial z_{u\tilde{f}k}} + \frac{\partial c^{(E)}_{\tilde{f}}\left(Z_{\tilde{f}}\right)}{\partial z_{u\tilde{f}k}} + \frac{\partial c_{\tilde{f}}(Z_{\tilde{f}})}{\partial z_{u\tilde{f}k}} - \alpha_k \frac{\partial q_k(Y, Z)}{\partial x_{u\tilde{f}k}} - \rho_k$$
$$+ \lambda^{1*} \frac{\partial c_{\tilde{f}}(Z_{\tilde{f}})}{\partial z_{u\tilde{f}k}} - \lambda_k^{2*} \frac{\partial q_k(Y, Z)}{\partial z_{u\tilde{f}k}},$$

the component F^4 is given by

$$F^4 = \overline{B} - \sum_{\tilde{f} \in \tilde{\mathcal{F}}_2} c_{\tilde{f}}\left(Z_{\tilde{f}}\right)$$

and, finally, the k-component, F^5_k, of $F^5(\mathcal{X})$ is given by

$$F^5_k = q_k(Y, Z) - Q_k,$$

and the feasible set $\mathcal{K}$ is defined as $\mathbb{K}$.

Since the feasible set $\mathcal{K}$ is convex and compact, to obtain an existence result for a solution to variational inequality (10), we have to impose that function F is a continuous function.

Following the classical theory of variational inequalities (see, for instance, [11]), we have the following existence result:

Theorem 2 (Existence) *If $\mathcal{K}$ is compact and convex and F is a continuous function on $\mathcal{K}$, then variational inequality* (10) *admits at least a solution.*

Moreover, we have the following uniqueness result:

Theorem 3 (Uniqueness) *Under the assumptions of Theorem 2, if the function $F(\mathcal{X})$ in (12) is strictly monotone on $\mathcal{K}$, that is:*

$$\langle (F(\mathcal{X}^1) - F(\mathcal{X}^2))^T, \mathcal{X}^1 - \mathcal{X}^2 \rangle > 0, \quad \forall \mathcal{X}^1,\ \mathcal{X}^2 \in \mathcal{K}, \mathcal{X}_1 \neq \mathcal{X}_2,$$

then variational inequality (12) or, equivalently, variational inequality (10), admits a unique solution.

4 An Illustrative Numerical Example

In this Section, we solve an illustrative numerical example to validate the effectiveness of the model. We consider G=5 users or devices on the ground, $U = 2$ controller UAVs, $\hat{F}_1 = 3$ pre-existing UAVs, $\tilde{F}_2 = 2$ additional UAVs and $K = 1$ type of services. The optimal solution are computed through the Euler Method (see [7] for a detailed description) using the Matlab program on an HP laptop with an AMD compute cores 2C+3G processor, 8 GB RAM. The numerical data and the size of the problem are constructed for easy interpretation purposes. We have the following example's data:

$$\rho_1 = 17,\ \alpha_1 = 5,\ Q_1 = 60, ;\ R_{11} = 10,\ R_{21} = R_{31} = 15,\ R_{41} = 12, R_{51} = 10;$$

$$\overline{S}_1 = 27,\ \overline{S}_2 = 37,\ s_1 = 2;\ S_1 = S_2 = S_3 = S_4 = S_5 = 30.$$

The parameters of the transmission cost of the service requests from users or devices to controller UAVs, from controller UAVs to any pre-existing UAV and any additional UAV functions are reported in Table 3 and we also assume:

$$\eta' = \kappa' = \mu' = 1;\ \delta = \epsilon = 0.3;\ \delta' = \epsilon' = 1.$$

Reminding that the cost due to add a new UAV at the highest level of the network is given by:

$$c_{\tilde{f}}(X^3_{\tilde{f}}) = \beta \left(\sum_{u=1}^{2} x_{u\tilde{f}1} \right)^2 + \beta' \sum_{u=1}^{2} x_{u\tilde{f}1}, \quad \forall \tilde{f} \in \tilde{\mathcal{F}}_2,$$

we conduct a sensitivity analysis, therefore we suppose 21 different scenarios, varying the parameters β and β' of the cost function for adding new UAVs, and the maximum available budget for additional UAVs, $\overline{B}$.

In Table 4, for each scenario, we reported: the main optimal solutions, that are $z^*_{u\tilde{f}k}$, the quantities of service requests sent by each controller UAV to the additional UAVs which the provider can decide to activate; c_{add}, the sum of the terms of the

Table 3 Used parameters of the transmission cost functions: η, κ and μ

	$g = 1$	$g = 2$	$g = 3$	$g = 4$	$g = 5$	$\hat{f} = 1$	$\hat{f} = 2$	$\hat{f} = 3$	$\tilde{f} = 1$	$\tilde{f} = 2$
$u = 1$	$\eta = 0.2$	$\eta = 0.2$	$\eta = 0.2$	$\eta = 0.3$	$\eta = 0.3$	$\kappa = 0.3$	$\kappa = 0.2$	$\kappa = 0.2$	$\mu = 0.3$	$\mu = 0.8$
$u = 2$	$\eta = 0.3$	$\eta = 0.3$	$\eta = 0.3$	$\eta = 0.2$	$\eta = 0.2$	$\kappa = 0.8$	$\kappa = 0.3$	$\kappa = 0.2$	$\mu = 0.2$	$\mu = 0.3$

Table 4 Optimal solutions ($z^*_{u\bar{f}k}$), additional costs (c_{add}) and computational times (t) for each scenario

Parameters	$\overline{B} = 60$						$\overline{B} = 45$						$\overline{B} = 30$					
	z_{111}	z_{121}	z_{211}	z_{221}	c_{add}	$t(s)$	z_{111}	z_{121}	z_{211}	z_{221}	c_{add}	$t(s)$	z_{111}	z_{121}	z_{211}	z_{221}	c_{add}	$t(s)$
$\beta = \beta' = 0$	6.32	3.46	9.3	9.2	0	0.72	6.32	3.46	9.3	9.2	0	0.59	6.32	3.46	9.3	9.2	0	0.63
$\beta = 0.5, \beta' = 0.1$	3.18	1.94	4.69	5.17	57.72	1.56	2.78	1.72	4.11	4.59	45	1.5	2.25	1.41	3.32	3.77	30	0.43
$\beta = 1, \beta' = 0.2$	2.12	1.34	3.12	3.56	53.5	0.97	1.93	1.23	2.85	3.27	45	0.89	1.56	1	2.3	2.68	30	1.45
$\beta = 2, \beta' = 0.4$	1.28	0.82	1.84	2.18	30.89	1.18	1.28	0.82	1.84	2.18	30.89	2.02	1.12	0.72	1.57	1.89	29.98	8.78
$\beta = 3, \beta' = 0.6$	0.91	0.59	1.29	1.56	30.93	2.9	0.91	0.59	1.29	1.56	30.93	6.07	0.9	0.58	1.27	1.53	29.98	9.42
$\beta = 5, \beta' = 1$	0.64	0.45	0.78	0.94	22.78	6.84	0.65	0.45	0.78	0.94	22.78	6.5	0.65	0.45	0.78	0.94	22.78	6.4
$\beta = 35, \beta' = 20$	0	0	0	0	0	8.48	0	0	0	0	0	9.7	0	0	0	0	0	8.81

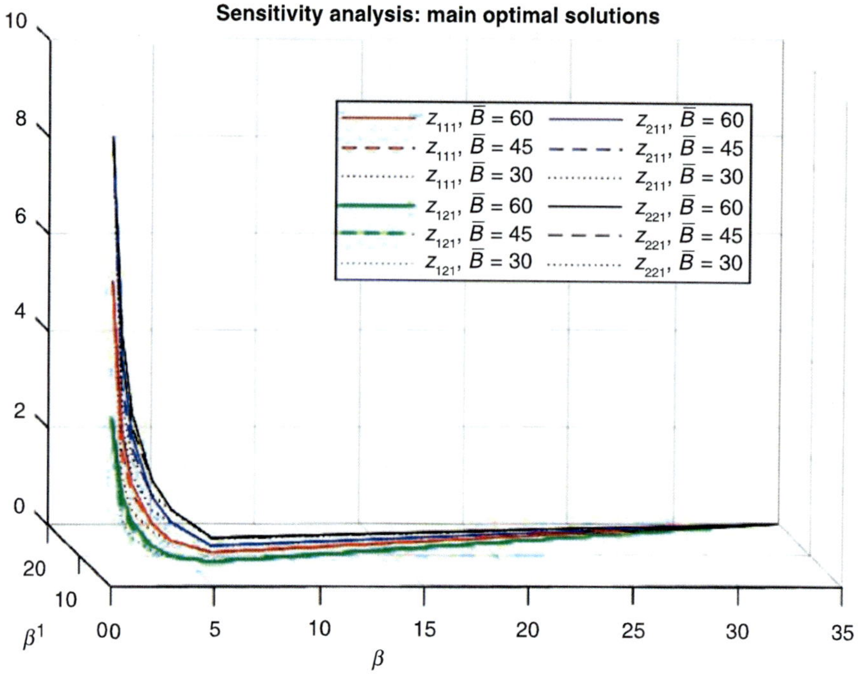

Fig. 2 Main optimal solutions $z_{u\tilde{f}k}$ under sensitivity analysis

objective function concerning the cost due to add new UAVs at the highest level of the network (that must be less than or equal to $\overline{B}$, the maximum available budget for adding new UAVs, as established by constraint (7)), and the computational time (in seconds) required to solve the scenario.

The optimal solutions, shown in Table 4 and in Fig. 2, clearly express a decreasing trend as the parameters of the cost functions increase. Particularly, at $\beta = \beta' = 0$, we obtain maximum values of $z_{u\tilde{f}k}$, while at $\beta = 35$ and $\beta' = 20$, we obtain that no additional UAVs are used because to add new UAVs is too expensive. Note that the trend of the curves is the same as the maximum budget varies and the budget constraint is always respected.

5 Conclusion

In this paper, we presented a supply chain network optimization model for the providing 5G network slices on demand to users and devices on the ground. We considered a three-tier supply chain consisting of a fleet of UAVs organized as a FANET, managed by the UAVs controllers, with the possibility to add additional UAVs to pre-existing ones. We proposed a system optimization perspective for the entire supply chain network in which services provider solves a maximization profit problem. We obtained a constrained optimization problem for which we derive the associated variational inequality formulation. Also, qualitative properties in terms of existence and uniqueness of the solution were provided. Finally, an illustrative numerical example was performed to validate the effectiveness of the model.

The model previously described can certainly be extended. In our future work, we are considering a closed-loop network in which users or devices on the ground receive the performed services, introducing the reliability level of the entire network and the related damage in case of services not performed. Furthermore, we are working on a more comprehensive model, in which we introduce a set of Time Slots, a bigger area to be covered and, therefore, a new heuristic to solve numerical examples on large instances.

Acknowledgments The research was partially supported by the research project "Programma ricerca di ateneo UNICT 2020-22 linea 2-OMNIA" of Catania. This support is gratefully acknowledged.

References

1. Amorosi, L., Chiaraviglio, L., Galan-Jimenez, J.: Optimal energy management of UAV-based cellular networks powered by solar panels and batteries: Formulation and solutions. IEEE Access **7**, 53698–53717 (2019)
2. Caruso, V., Daniele, P.: A network model for minimizing the total organ transplant costs. Eur. J. Oper. Res. **266**(2), 652–662 (2018)
3. Chiaraviglio, L., Amorosi, L., Blefari-Melazzi, N., Dell'Olmo, P., Lo Mastro, A., Natalino, C., Monti, P.: Minimum cost design of cellular networks in rural areas with UAVs, optical rings, solar panels, and batteries. IEEE Trans. Green Commun. Netw. **3**(4), 901–918 (2019)
4. Colajanni, G., Daniele, P.: A mathematical network model and a solution algorithm for IaaS cloud computing. Netw. Spat. Econ., 1–21 (2019)
5. Colajanni, G., Daniele, P., Sciacca, D.: A projected dynamic system associated with a cybersecurity investment model with budget constraints and fixed demands. J. Nonlinear Var. Anal. **4**(1), 45–61 (2019/2020). Available online at http://jnva.biemdas.com. 10.23952/jnva.4.2020.1.05
6. Daniele, P., Sciacca D.: An optimization model for the management of green areas. Intl. Trans. Oper. Res. **0**, 1–23 (2021)
7. Dupuis, P., Nagurney, A.: Dynamical systems and variational inequalities. Ann. Oper. Res. **44**, 9–42 (1993)
8. Galluccio, L., Grasso, C., Grasso, M., Raftopoulos, R., Schembra, G.: Measuring QoS and QoE for a Softwarized Video Surveillance System in a 5G Network. In: 2019 IEEE International Symposium on Measurements & Networking (M&N), pp. 1–6. IEEE (2019)
9. Grasso, C., Schembra, G.: A fleet of MEC UAVs to extend a 5G network slice for video monitoring with low-latency constraints. J. Sens. Actuator Netw. **8**(1), 3 (2019)
10. Khan, M.F., Yau, K.L.A.: Route selection in 5G-based flying ad-hoc networks using reinforcement learning. In: 2020 10th IEEE International Conference on Control System, Computing and Engineering (ICCSCE), pp. 23–28. IEEE (2020)

11. Nagurney, A.: Network Economics: A Variational Inequality Approach. Kluwer Academic Publishers, Boston, MA (1993)
12. Nagurney, A.: Supply Chain Network Economics: Dynamics of Prices, Flows, and Profits. Edward Elgar Publishing, Cheltenham, England (2006)
13. Nagurney, A., Daniele, P., Shukla, S.: A supply chain network game theory model of cybersecurity investments with nonlinear budget constraints. Ann. Oper. Res. **248**(1), 405–427 (2017)

A MIP Model for Freight Consolidation in Road Transportation Considering Outsourced Fleet

Thiago Vieira and Pedro Munari

Abstract We address the freight consolidation problem under the context of road transportation with outsourced fleet, motivated by the real-life situation faced by a major manufacturer of school supplies located in Brazil. Given a set of shipments scheduled for the next few days, and a set of available vehicles at the carriers, the company has to determine how to best assign shipments to vehicles in a way to minimize the total transportation cost. This is a challenging case of vehicle consolidation, in which each carrier has a complex pricing table that follows particular rules, rates and taxes. Prices are defined according to the vehicle type (heterogeneous fleet), number of deliveries (visits to redispatching points) and the individual price and weight of items in the shipments consolidated in the truck. These components cause a piecewise linear behavior of the cost function, which makes the consolidation even more difficult. To aid this decision-making process, we propose a mixed-integer linear programming (MIP) model that fully represents the problem. We are not aware of any other model or solution strategy that includes all the features observed in the addressed situation. The results of computational experiments using real-life instances provided by the manufacturer show the benefits of using the proposed model in practice, as we observed reductions of more than 44% in comparison to the freight consolidation policy of the company.

Keywords Freight consolidation · Piecewise linear function · Mixed-integer linear optimization

1 Introduction

Freight (or shipment) consolidation is an essential practice in logistics operations, being widely used in transportation modes [17]. It consists of combining different items, produced and used at different locations and times with the same destination,

T. Vieira (✉) · P. Munari
Department of Production Engineering, Federal University of São Carlos, São Carlos, Brazil
e-mail: munari@dep.ufscar.br

A. Masone et al. (eds.), *Optimization and Data Science: Trends and Applications*, AIRO Springer Series 6, https://doi.org/10.1007/978-3-030-86286-2_8

into single container or truck loads, in a way to reduce transportation expenditures by taking advantage of economies of scale, better organize the deliveries, minimize damages to the customers and provide greater inventory control [4, 9].

Freight consolidation can be classified based on the supply chain planning hierarchy, mainly into strategic and operational levels [15]. The strategic level concerns the design of distribution networks, involving definitions of factors such as the quantity, the location and the processing capacity of each distribution center (DC), with the aim of minimizing the long-term overall costs [1, 5, 12]. On the other hand, operational level deals with the planning and execution of goods distribution considering an existing network configuration. [18] presented a model to minimize costs and carbon dioxide emissions in long-haul transportation by freight consolidation. [13] proposed a look-ahead heuristic that achieves economy of scale by shipping larger quantities in a system with stochastic demand and a single consolidation point near the suppliers. [16] introduced a nonlinear optimization model to coordinate shipments between suppliers and customers through a DC. Other freight consolidation models at the operational level can be found in [3, 14] and [6].

In a more practical perspective, [10] and [11] classify the consolidation strategies according to three types: inventory (or temporal), vehicle, terminal. Inventory consolidation creates a stock of items by holding shipments until reaching a minimum load size. In vehicle consolidation, small load shipments are picked up and dropped off along a multi-stop route by the same vehicle, in a way that the combined big loads maximize the utilization of the resource. Finally, terminal consolidation collects items from different origins to a transshipment center and group them into larger shipments based on their destinations (e.g., break-bulk or cross-docking practices).

In this paper, we address freight consolidation at operational level, under vehicle and temporal strategies, for effective road transportation using outsourced fleet. We consider the real-life decision-making process faced by the logistics department of a manufacturer of school supplies located in Brazil. In short, in the company's operation, the cargo is shipped from their DC by *collection* carriers, hired by the company, and then proceeds to the DCs of *redispatch* carriers, which are hired by the customers and are fully responsible for delivering the goods to them. Given the orders scheduled for the next days and the available fleet of the collection carriers, the company has to decide how to best consolidate these orders and assign them to vehicles, in a way to minimize the total transportation cost from its DC to the DCs of the *redispatch* carriers. The difficulty in this decision making process comes from the large number of orders and the big variety of vehicles and carriers, and from the limited processing capacity at the DCs. Additionally, the collection carriers have complex pricing tables, based on vehicle types, number of deliveries of a vehicle, and the individual price and weight of items. This causes a nonlinear structure in the total transportation cost, which brings additional challenges for modeling and solving the problem.

We propose a mixed-integer linear programming (MIP) model for minimizing the total transportation cost in the scope of the addressed situation, considering an extension of the Variable Sized Bin-Packing Problem (VSBPP) [8], which is an NP-hard combinatorial optimization problem that consists of determining the optimal arrangement of smaller units (items) within larger units (bins) [7]. The proposed model and the analyses presented in this paper can benefit companies with a similar situation, as well as be extended to other situations related to freight consolidation using outsourced fleet and a nonlinear cost structure.

The remainder of this paper is structured as follows. In Sect. 2 we describe the problem and in Sect. 3 we introduce a new MIP formulation. In Sect. 4, we show the results of computational experiments using real-life data and, lastly, in Sect. 5 we present our conclusions and the next steps.

2 Problem Description

As mentioned before, we consider the case of a company that operates in the printing industry, specifically in the stationery segment (school, office, calendar and household supplies), located in São Paulo, Brazil. The final products are stored in the company's DC, where freight consolidation and other inventory management activities are fulfilled. Then, collection carriers transfer loads to the DCs of redispatch carriers, which are then finally delivered to the customers. The manufacturer covers the transportation expenses only between its DC and the DCs of the redispatch carriers and, hence, it is a Free on Board (FOB) redispatch operation. The processing capacity at the company's DC regarding the collection of goods must not exceed 90 ton/day. The redispatch carriers also impose a processing capacity and, thus, the company assumes the standard limitation of 36 ton/day per carrier.

The addressed problem can be seen as an extension of the VSBPP, in which carriers' vehicles correspond to the bins, items refer to the goods to be dispatched (represented by orders), and the utility of each bin is determined by a piecewise linear function that depends on the items allocated to the bin. This piecewise linear function results from the rules used by the collection carriers in their pricing tables. They are established by contracts between the company and the carriers and cannot be easily changed.

The pricing table defines the transportation cost of each order assigned to a vehicle, based on the individual cost and weight of the goods and on the total number of visits to the DCs of redispatch carriers. Thus, the transportation cost of an order depends not only on its attributes (cost and weight) but also on the number of visits of the vehicle to which the order is assigned to. Therefore, the consolidation must take into account that the larger the number of visits made by a vehicle, the larger the carrier charges will be, for each order assigned to that vehicle.

Table 1 illustrates how a pricing table is often structured. In this example, there are two available vehicle types (*Box Truck 14-ton* and *Semitrailer 24-ton*), both offered by the same collection carrier. Each vehicle type has a price range that depends on the number of deliveries, and this range is defined by the bounds DL and DU. A price range can be composed of subranges based on the weight of items

Table 1 Example of the pricing table of a collection carrier

Vehicle type	Price range	DL	DU	Price subrange	WL	WU	$Cost^A$	$Cost^B$	...
Box Truck 14-ton	7	1	6	1	0	∞	**0.13**	57.22	...
	8	7	∞	1	0	∞	**0.18**	55.00	...
Semitrailer 24-ton	9	1	6	1	0	∞	**0.15**	37.61	...
	10	7	20	1	0	50	26.95	57.22	...
				2	51	100	32.60	55.00	...
				3	101	∞	**0.24**	57.22	...

in an order (defined by *WL* and *WU*), and for each subrange there is an associated price ($Cost^A, Cost^B, \ldots$). Let us assume that the values in bold in column $Cost^A$ are rates based on the weight of an order, whereas the remaining values in this column and in column $Cost^B$ are fixed costs (i.e., they do not depend on weights). This definition of ranges and subranges cause the piecewise linear behavior of the transportation cost. For example, if we assign a vehicle of type *Box Truck 14-ton* to execute a route with 5 deliveries, it falls in the price range 7 (1 to 6 visits). Then, the cost applied to each order assigned to this vehicle is $0.13w_i + 57.22 + \ldots$, where w_i is the weight of order i. Observe that in this price range, w_i is unbounded (as $WU = \infty$). However, if we assign a vehicle type *Semitrailer 24-ton* to a route in which the number of deliveries falls between 7 and 20 (price range 10), then the cost of an order is computed based on its total weight: if it has 50 Kg or less, then we apply the cost $26.95 + 57.22 + \ldots$; whereas if it has more than 50 Kg, but not more than 100 Kg, we apply $32.60 + 55.00 + \ldots$; and finally, if it has more than 100 Kg, we apply $0.24w_i + 57.22 + \ldots$.

To effectively include these costs in the MIP model, we propose in the next section a way to linearize the transportation cost function. This is done by incorporating one additional index into variables and parameters, to explicitly state the price subrange that is chosen for a vehicle. Finally, to avoid the risk of a possible reformulation of the contract, disinterest in the provision of service by any carrier, or even for the total break of the partnership, the company has to ensure a minimum occupancy volume factor for each vehicle, i.e., $OC = loading/capacity$.

3 Mathematical Formulation

Consider a set of n orders, denoted as $\mathcal{N} = \{1, \ldots, n\}$, and the set of time periods $\mathcal{T}$ representing the time horizon considered in the problem. The orders have to be dispatched from the company's DC to the DCs of m redispatch carriers denoted by the set $\mathcal{M} = \{1, \ldots, m\}$. Let $\mathcal{V}$ be the set of available vehicles at the collection carriers. We define $\mathcal{I}_j \subset \mathcal{N}$ as the subset of orders whose destination is the DC of the redispatch carrier $j \in \mathcal{M}$. Additionally, let $\mathcal{F}_v$ be the set of price ranges that applies for vehicle $v \in \mathcal{V}$, according to the pricing table of the collection carrier that owns this vehicle. To guarantee a satisfactory occupation level in the vehicles, we allow orders to be rejected in a solution of the model, as an artifice to postpone such orders to a subsequent planning (e.g., a next run of the model, using a forward time horizon). To prevent the rejection of orders with short due dates, we define the subset $NR \subset \mathcal{N}$ of orders that cannot be rejected in a solution of the model. We further define the following input parameters:

- w_i: weight of order i (in kilograms);
- TC_{if}: total unit transportation cost for order i and price range f;
- $PCcol$: processing capacity of company's DC in Kg/day;
- $PCred_j$: processing capacity of the redispatch carrier j in Kg/day;
- WLC_v: weight capacity of vehicle v in (in kilograms);
- OC: minimum occupancy volume factor of the vehicles, given in percentage;
- DL_f: minimum number of deliveries allowed when choosing price range f;
- DU_f: maximum number of deliveries allowed when choosing price range f;
- Due_i: maximum period for shipping the order i.

Since consolidation represents the total loading assigned to a vehicle, according to a certain price range, the following decision variables are defined:

- $x_{ifvt} = \begin{cases} 1, & \text{if order } i \text{ is assigned to price range } f \text{ of vehicle } v \text{ in period } t; \\ 0, & \text{otherwise.} \end{cases}$
- $u_{fvt} = \begin{cases} 1, & \text{if price range } f \text{ of vehicle } v \text{ in period } t \text{ is activated;} \\ 0, & \text{otherwise.} \end{cases}$
- $z_{jfvt} = \begin{cases} 1, & \text{if the redispatch carrier } j \text{ is visited by vehicle } v \text{ with price} \\ & \text{range } f \text{ in period } t; \\ 0, & \text{otherwise.} \end{cases}$
- s_i: continuous variable that assumes the value of 1 if order i is not allocated to any vehicle (allowing postponement of orders for subsequent planning), and 0 otherwise.

The optimization criterion of this approach seeks to minimize the total transportation cost. Moreover, we define $TCmax_i = 2 \max_f \{TC_{if}\}$ in a way to penalize each rejected order. This definition is an artificial cost that, in practice, should be estimated empirically. Using the defined parameters and decision variables, we state the following MIP model for the addressed freight consolidation problem:

$$\min \sum_{t \in \mathcal{T}} \sum_{v \in \mathcal{V}} \sum_{f \in \mathcal{F}_v} \sum_{i \in \mathcal{N}} TC_{if} x_{ifvt} + \sum_{i \in \mathcal{N}} TCmax_i s_i \tag{3.1}$$

$$\text{s.t.} \sum_{t \in \mathcal{T}} \sum_{v \in \mathcal{V}} \sum_{f \in \mathcal{F}_v} x_{ifvt} + s_i = 1; \ \forall\, i \in \mathcal{N}; \tag{3.2}$$

$$\sum_{f \in \mathcal{F}_v} u_{fvt} \leq 1; \ \forall \, v \in \mathcal{V}; t \in \mathcal{T}; \tag{3.3}$$

$$\sum_{v \in \mathcal{V}} \sum_{f \in \mathcal{F}_v} \sum_{i \in \mathcal{N}} w_i x_{ifvt} \leq PCcol; \ \forall \, t \in \mathcal{T}; \tag{3.4}$$

$$\sum_{v \in \mathcal{V}} \sum_{f \in \mathcal{F}_v} \sum_{i \in \mathcal{I}_j} w_i x_{ifvt} \leq PCred_j; \ \forall \, j \in \mathcal{M}; t \in \mathcal{T}; \tag{3.5}$$

$$(OC.WLC_v)u_{fvt} \leq \sum_{i \in \mathcal{N}} w_i x_{ifvt} \leq WLC_v u_{fvt};$$

$$\forall \, f \in \mathcal{F}_v; v \in \mathcal{V}; t \in \mathcal{T}; \tag{3.6}$$

$$x_{ifvt} \leq z_{jfvt}; \ \forall \, i \in \mathcal{I}_j; \, j \in \mathcal{M}; \, f \in \mathcal{F}_v; v \in \mathcal{V}; t \in \mathcal{T}; \tag{3.7}$$

$$z_{jfvt} \leq \sum_{i \in \mathcal{I}_j} x_{ifvt}; \ \forall \, j \in \mathcal{M}; \, f \in \mathcal{F}_v; v \in \mathcal{V}; t \in \mathcal{T}; \tag{3.8}$$

$$DL_f u_{fvt} \leq \sum_{j \in \mathcal{M}} z_{jfvt} \leq DU_f; \ \forall \, f \in \mathcal{F}_v; v \in \mathcal{V}; t \in \mathcal{T}; \tag{3.9}$$

$$\sum_{v \in \mathcal{V}} \sum_{f \in \mathcal{F}_v} \sum_{i \in \mathcal{N}} w_i x_{ifv,t-1} \geq \sum_{v \in \mathcal{V}} \sum_{f \in \mathcal{F}_v} \sum_{i \in \mathcal{N}} w_i x_{ifvt}; \ \forall \, t \in \mathcal{T} \mid t > 1; \tag{3.10}$$

$$x_{ifvt} = 0; \ \forall \, i \in \mathcal{N}; \, f \in \mathcal{F}_v; v \in \mathcal{V}; \ t \in \mathcal{T} \mid t > Due_i; \tag{3.11}$$

$$s_i = 0; \ \forall \, i \in \mathcal{N} \mid i \in NR; \tag{3.12}$$

$$x_{ifvt} \in \{0, 1\}; \ \forall \, i \in \mathcal{N}; \, f \in \mathcal{F}_v; v \in \mathcal{V}; t \in \mathcal{T}; \tag{3.13}$$

$$u_{fvt} \in \{0, 1\}; \ \forall \, f \in \mathcal{F}_v; v \in \mathcal{V}; t \in \mathcal{T}; \tag{3.14}$$

$$z_{jfvt} \in \{0, 1\}; \ \forall \, j \in \mathcal{M}; \, f \in \mathcal{F}_v; v \in \mathcal{V}; t \in \mathcal{T}; \tag{3.15}$$

$$s_i \geq 0; \ \forall \, i \in \mathcal{N}. \tag{3.16}$$

The objective function (3.1) represents the trade-off between minimizing the total transportation cost and maximizing of the occupancy volume of a vehicle by penalizing orders rejections. If the problem was originally feasible in relation to a complete assignment of orders to vehicles, in an optimal solution we have $\sum_{i=1}^{n} TCmax_i s_i = 0$. As mentioned before, a feasible solution with rejected orders means that these orders were not allocated to a vehicle in this particular solution, and hence they should be included in a future planning, in another execution of the model. Constraints (3.2) ensure that an order i is allocated to at most one price range f of vehicle v in time period t. The use of s_i enables the rejection of an order if the satisfactory occupation of the vehicles cannot be achieved, meaning the this order will be considered in a subsequent run of the model. Constraints (3.3) guarantee that only a single price range f can be associated with each vehicle v in period t, activating variable u_{fvt} if $\sum_{i \in \mathcal{N}} x_{ifvt} > 0$. Constraints (3.4) and (3.5) enforce the processing capacities of the company's DC (90 ton/day) and of the DCs of redispatch carriers (36 ton/day per carrier), and constraints (3.6) enforce the minimum occupation and maximum capacity of each vehicle v. Constraints (3.7)–(3.9) concern the number of visits of a vehicle, given by the number of different redispatch carriers in a consolidation, where constraints (3.7) and (3.8) ensure that $\sum_{i \in \mathcal{I}_j} x_{ifvt} > 0$ if, and only if, $z_{jfvt} = 1$, whereas constraints (3.9) relate the number of visits with the price ranges. Constraints (3.10) prevent solution symmetry in the model, as there is no difference in the assignment of orders in relation to periods of time. Constraints (3.11) and (3.12) guarantee that no order is allocated after the delivery due date and that orders that cannot be rejected must be fulfilled in the planning in question, respectively. Finally, constraints (3.13)–(3.16) impose the domain of decision variables. Note that we do not need to define variable s_i as binary, as constraints (3.2) and (3.13) preserve its domain in a feasible solution. The usage of the index f to differentiate the price ranges and of variable u_{fvt} in the bounds of constraints (3.6) and (3.9), allowed us to completely linearize the transportation cost function that was originally piecewise linear [2, 19].

4 Computational Experiments

The proposed MIP model was coded in C++ language, using the API Concert Technology of the IBM CPLEX Optimization Studio v.12.9. We ran computational experiments using data provided by the company, based on their operation history of a month with high demand of goods (September). We created three independent classes of instances, named hereafter as Classes 1, 3 and 5. Instances in Class 1 are defined by orders dispatched on a single day, resulting in 30 different instances in which $\mathcal{T} = \{1\}$; instances in Class 3 comprises orders of three consecutive days of operation and thus we obtained 10 instances with $\mathcal{T} = \{1, 2, 3\}$; and Class 5 has six instances with $\mathcal{T} = \{1, \ldots, 5\}$. We considered a total of 25 available vehicles ($|\mathcal{V}|$) from three different collection carriers, and an occupancy volume factor $OC = 90\%$. All experiments were performed on a PC with an Intel(R) Core i7 4790K 3.6 GHz CPU and 16 GB of RAM. We imposed a time limit of 1 hour and kept the default tolerance of CPLEX for the relative optimality gap (0.01%).

Table 2 presents the results of solving the proposed model using the general-purpose MIP solver of CPLEX. The first two columns show the instance name and its total number of orders ($|\mathcal{N}|$). The third column presents the actual company's transportation cost to dispatch the accepted orders in the instance ($Cost_{nA}$). The remaining nine columns refer to the solution obtained by the model and show the number of accepted orders in the solution (nA), number of rejected orders (nR), the lower (OF_{lb}) and upper (OF_{ub}) bounds, the relative gap computed as $100\%.(OF_{ub}$-$OF_{lb})/(OF_{ub}+10^{-10})$, objective function value for accepted orders of the instance (OF_{nA}), the percentage of cost reduction in the objective function with respected to the company's cost (RD), computed as $100\%.(OF_{nA}$-$Cost_{nA})/Cost_{nA}$, computational time in seconds (CPUt) and the average vehicle's occupancy volume

Table 2 Computational results of the proposed model

Instance		Company	Optimization model								
Name	$\|\mathcal{N}\|$	$Cost_{nA}$	nA	nR	OF_{lb}	OF_{ub}	Gap (%)	OF_{nA}	RD (%)	CPUt	AOC (%)
09_01	105	35,775.10	105	0	16,348.05	16,349.08	0.006	16,349.08	−54.30	66.10	97.67
09_02	208	33,229.18	208	0	20,135.05	20,158.90	0.118	20,158.90	−39.33	3,602.63	95.68
09_03	272	43,137.12	272	0	24,232.34	24,234.77	0.010	24,234.77	−43.82	260.67	92.35
09_04	164	31,130.22	164	0	18,436.33	18,437.91	0.009	18,437.91	−40.77	248.42	97.33
09_05	215	39,622.24	215	0	19,968.48	19,969.91	0.007	19,969.91	−49.60	112.30	92.61
09_06	112	25,678.10	112	0	16,205.72	16,206.68	0.006	16,206.68	−36.89	130.35	97.25
09_07	119	29,425.97	119	0	16,710.07	16,711.30	0.007	16,711.30	−43.21	125.45	97.48
09_08	111	29,483.12	111	0	16,127.74	16,129.36	0.010	16,129.36	−45.29	187.88	95.52
09_09	145	46,784.97	145	0	17,446.33	17,446.72	0.002	17,446.72	−62.71	218.19	91.17
09_10	132	27,307.04	132	0	16,262.81	16,264.21	0.009	16,264.21	−40.44	437.14	95.00
09_11	93	29,949.03	93	0	15,126.45	15,127.89	0.010	15,127.89	−49.49	159.53	97.05
09_12	118	28,102.72	118	0	16,180.90	16,182.48	0.010	16,182.48	−42.42	294.97	96.13
09_13	62	17,915.87	62	0	13,750.60	13,751.64	0.008	13,751.64	−23.24	19.29	97.38
09_14	143	31,087.49	143	0	18,348.04	18,349.88	0.010	18,349.88	−40.97	144.19	94.49
09_15	188	37,873.69	188	0	19,197.12	19,199.03	0.010	19,199.03	−49.31	94.52	95.73
09_16	149	31,886.92	149	0	17,816.87	17,818.65	0.010	17,818.65	−44.12	598.70	95.58
09_17	137	27,315.95	137	0	16,819.37	16,821.03	0.010	16,821.03	−38.42	48.06	97.41
09_18	152	30,910.19	152	0	17,704.28	17,704.78	0.003	17,704.78	−42.72	16.94	95.69
09_19	131	28,485.88	131	0	16,989.17	17,032.77	0.256	17,032.77	−40.21	3,600.77	94.31
09_20	110	25,931.25	110	0	16,142.96	16,144.58	0.010	16,144.58	−37.74	128.82	98.65
09_21	198	56,898.09	198	0	19,102.42	19,104.31	0.010	19,104.31	−66.42	96.27	95.61
09_22	197	36,528.67	197	0	19,485.18	19,487.13	0.010	19,487.13	−46.65	537.24	95.72
09_23	138	31,228.36	138	0	17,394.85	17,396.59	0.010	17,396.59	−44.29	122.18	94.74

(continued)

09_24	100	22,787.24	100	0	15,181.43	15,182.89	0.010	15,182.89	−33.37	262.75	94.83
09_25	153	39,334.95	153	0	17,620.65	17,622.40	0.010	17,622.40	−55.20	116.27	98.10
09_26	111	29,124.36	111	0	16,508.78	16,510.43	0.010	16,510.43	−43.31	46.01	94.65
09_27	105	27,354.90	105	0	16,347.13	16,348.76	0.010	16,348.76	−40.23	223.97	95.21
09_28	112	28,042.07	112	0	16,067.18	16,068.78	0.010	16,068.78	−42.70	1,837.88	97.19
09_29	116	26,944.13	116	0	16,481.21	16,482.79	0.010	16,482.79	−38.83	128.28	98.76
09_30	173	33,835.18	173	0	18,408.10	18,409.82	0.009	18,409.82	−45.59	24.91	95.28
Avg	**142.3**	**32,103.67**	**142.3**	**0**	**17,418.19**	**17,421.85**	**0.021**	**17,421.85**	**−44.05**	**463.02**	**95.82**
09_01-03	585	112,141.40	580	0	53,659.20	58,502.94	8.279	58,502.94	−47.83	3,601.17	95.57
09_04-06	491	96,430.56	491	0	49,125.70	54,586.87	10.005	54,586.87	−43.39	3,602.82	94.02
09_07-09	375	105,694.06	375	0	46,380.56	47,652.74	2.670	47,652.74	−54.91	3,602.01	96.79
09_10-12	343	85,358.79	343	0	44,789.29	45,790.40	2.186	45,790.40	−46.36	3,601.91	96.77
09_13-15	393	86,877.05	393	0	46,225.43	47,831.12	3.357	47,831.12	−44.94	3,602.49	96.55
09_16-18	438	90,113.06	438	0	47,665.25	51,370.45	7.213	51,370.45	−42.99	3,604.96	96.56
09_19-21	439	111,315.22	439	0	47,734.06	53,403.20	10.616	53,403.20	−52.03	3,602.46	95.82
09_22-24	435	90,544.27	435	0	46,969.37	49,168.21	4.472	49,168.21	−45.70	3,602.73	96.48
09_25-27	369	95,814.21	369	0	45,693.03	49,297.49	7.312	49,297.49	−48.55	3,602.32	93.94
09_28-30	401	88,821.38	401	0	46,995.67	50,017.25	6.041	50,017.25	−43.69	3,606.07	97.76
Avg	**426.9**	**96,311.00**	**426.4**	**0**	**47,523.76**	**50,762.07**	**6.215**	**50,762.07**	**−47.04**	**3,602.89**	**96.03**
09_01-05	964	114,992.44	512	452	86,787.15	246,297.73	64.763	64,524.29	−43.89	3,601.00	97.56
09_06-10	619	123,988.01	412	207	73,856.77	157,017.04	52.963	61,053.14	−50.76	3,600.63	94.50
09_11-15	604	120,183.53	473	131	73,684.87	9.77E+07	99.925	66,234.27	−44.89	3,604.97	95.52
09_16-20	679	113,668.80	500	179	75,831.27	1.42E+08	99.947	62,474.83	−45.04	3,605.45	94.09
09_21-25	786	132,915.59	439	347	79,712.96	208,061.22	61.688	57,446.64	−56.78	3,604.91	98.45
09_26-30	617	117,909.96	448	169	73,946.11	1.29E+08	99.943	62,078.60	−47.35	3,604.95	95.54
Avg	**711.5**	**120,609.72**	**464**	**247.5**	**77,303.19**	**6.16E+07**	**79.871**	**62,301.96**	**−48.12**	**3,603.65**	**95.94**

(AOC). In addition to the results for each instance, Table 2 shows at the end of each class the average values for the instances in the class. It is worth mentioning that the rejected orders are those not allocated to a vehicle in the solution obtained for the instance. This means that the solver could not find a solution in which all orders are assigned to vehicles and, therefore, these orders are not fulfilled in this solution. In our experiments, this situation appears in solutions with large optimality gaps only, as the solver was not able to find a relatively good solution when the time limit was reached. For solutions with small gaps, we do not see any rejection, thus all orders are fulfilled in the time horizon of the instance.

For instances in Class 1, on average, we observe a reduction in the transportation costs of 44.05%, with an execution time of 463.02 seconds and an occupation of 95.82%. Also, the solution obtained using the model resulted in a reduction greater than 60% for two instances, namely *09_09* and *09_21*. There were no rejected orders in the instances of this class. Only in two instances CPLEX achieved the time limit of one hour.

In Class 3, only two instances had a gap below 3%. Although the average gap was 6.22%, there were no rejected orders and the economy was 47.04% with an occupancy factor of 96.03%. Note that CPLEX reached the time limit in all instances in this class. Finally, in Class 5, the difficulty significantly increased, as the average gap was greater than 50% when CPLEX reached the time limit, and the average number of rejected orders was 247.5 (34.78% of the total), given that these solutions have a large optimality gap. Despite that, the corresponding savings were high (48.12%, on average), and the average occupancy was around 95.94%.

In summary, we notice that the computational difficulty of solving the model with a larger number of time periods increased significantly. Nevertheless, the solutions resulted in consolidations that were considerably less costly than the consolidation carried out by the company.

5 Concluding Remarks

We addressed a freight consolidation problem in road transportation with outsourced fleet and FOB freight, considering the real-life case of a manufacturer of school supplies. We proposed a mixed-integer linear optimization model that seeks to minimizing the total cost of transportation while satisfying all requirements that are considered relevant by the company. Due to the complex structure of the pricing tables practiced by the carriers, the transportation cost function has a piecewise linear structure, which was effectively linearized in the model. The results obtained by computational experiments with data provided by the company showed average savings of more than 44% with respect to the history of the company, indicating that the model has the potential to improve the addressed decision making process. An interesting next step for this study is to develop MIP-based heuristics using the proposed formulation, in an attempt to solve instances with larger number of time periods. Another research topic would be the incorporation of uncertainties into the model, through stochastic programming or robust optimization.

Acknowledgments The authors are thankful to the company involved in this study and for the financial support provided by the Coordenação de Aperfeiçoamento de Pessoal de Nível Superior—Brasil (CAPES) [Finance code 001], the National Council for Scientific and Technological Development (CNPq) [grant number 313220/2020-4], and the São Paulo Research Foundation (FAPESP) [grant numbers 20/11602-5, 19/23596-2 and 16/01860-1].

References

1. Ahuja, R., Magnanti, T.L., Orlin, J.B.: Network Flows: Theory, Algorithms and Applications. Rentice-Hall, New Jersey (1993)
2. Croxton, K.L., Gendron, B., Magnanti, T.L.: A comparison of mixed-integer programming models for nonconvex piecewise linear cost minimization problems. Management Science **49**(9), 1268–1273 (2003a)
3. Croxton, K.L., Gendron, B., Magnanti, T.L.: Models and methods for merge-in-transit operations. Transportation Science **37**(1), 1–22 (2003b)
4. Cruz, C.A., Munari, P., Morabito, R.: A branch-and-price method for the vehicle allocation problem. Comput. Ind. Eng. **149**, 106745 (2020)
5. Cunha, C.B., Silva, M.R.: A genetic algorithm for the problem of configuring a hub-and-spoke network for a ltl trucking company in brazil. Eur. J. Oper. Res. **179**(3), 747–758 (2007)
6. Dror, M., Hartman, B.C.: Shipment consolidation: Who pays for it and how much? Management Science **53**(1), 78–87 (2007)
7. Dyckhoff, H., Finke, U.: Cutting and Packing in Production and Distribution: A Typology and Bibliography. Springer Science & Business Media (2012)
8. Friesen, D.K., Langston, M.A.: Variable sized bin packing. SIAM J. Comput. **15**(1), 222–230 (1986)
9. Ghiani, G., Laporte, G., Musmanno, R.: Introduction to Logistics Systems Planning and Control. Wiley (2004)
10. Hall, R.W.: Consolidation strategy: inventory, vehicles and terminals. J. Bus. Logist. **8**(2), 57 (1987)
11. Higginson, J., Bookbinder, J.H.: Policy recommendations for a shipment-consolidation program. J. Bus. Logist. **15**(1), (1994)
12. Martín, J.C., Román, C.: Analyzing competition for hub location in intercontinental aviation markets. Transport. Res. E Logist. Transport. Rev. **40**(2), 135–150 (2004)
13. Nguyen, C., Dessouky, M., Toriello, A.: Consolidation strategies for the delivery of perishable products. Transport. Res. E Logist. Transport. Rev. **69**, 108–121 (2014)
14. Popken, D.A.: An algorithm for the multiattribute, multicommodity flow problem with freight consolidation and inventory costs. Operations Research **42**(2), 274–286 (1994)
15. Qin, H., Zhang, Z., Qi, Z., Lim, A.: The freight consolidation and containerization problem. Eur. J. Oper. Res. **234**(1), 37–48 (2014)
16. Song, H., Hsu, V.N., Cheung, R.K.: Distribution coordination between suppliers and customers with a consolidation center. Operations Research **56**(5), 1264–1277 (2008)
17. Tyan, J.C., Wang, F.K., Du, T.C.: An evaluation of freight consolidation policies in global third party logistics. Omega **31**(1), 55–62 (2003)
18. Ülkü, M.A.: Dare to care: Shipment consolidation reduces not only costs, but also environmental damage. Int. J. Prod. Econ. **139**(2), 438–446 (2012)
19. Vielma, J.P., Ahmed, S., Nemhauser, G.: Mixed-integer models for nonseparable piecewise-linear optimization: Unifying framework and extensions. Operations Research **58**(2), 303–315 (2010)

Part III
Logistics

Energy-Oriented Inter-Vehicle Distance Optimization for Heterogeneous E-Platoons

Bianca Caiazzo, Angelo Coppola, Alberto Petrillo, and Stefania Santini

Abstract The development of Connected and Autonomous Vehicles (CAVs) has the potential to improve the energy efficiency of the transportation system. Since the inter-vehicle distance plays a crucial role for energy-saving purposes, this paper proposes a novel optimization algorithm to compute the optimal gap distance in a heterogeneous platoon of electric CAVs by exploiting a distance-dependent air drag coefficient formulation. Specifically, the proposed method exploits the leader speed and acceleration profiles information, as well as road slope one, to compute the optimal inter-vehicle distance w.r.t. the preceding vehicle so as to reduce the air-drag coefficient and, hence, reduce energy consumption. The proposed algorithm exploits Nonlinear Programming method, taking into account safety constraints in order to avoid collisions among vehicles. The effectiveness of the approach is evaluated via the Matlab/Simulink simulation platform by considering two driving scenarios, namely *base* scenario, where the Constant Time Headway spacing policy is adopted, and *optimized* scenario, where the proposed algorithm is exploited to compute the optimal inter-vehicle distances. Numerical results confirm the effectiveness of the proposed optimization strategy in reducing energy consumption w.r.t. base scenario.

Keywords Energy efficiency · Nonlinear programming · Spacing policy · Heterogeneous platoon

The authors are listed in alphabetic order.

This work was supported by the Regione Campania, Italy, and FCA Group, through P.O.R. CAMPANIA FSE 2014/2020 CUP E66C18000900002.

B. Caiazzo · A. Coppola (✉) · A. Petrillo · S. Santini
The authors are with the Department of Electrical Engineering and Information Technology (DIETI), University of Naples Federico II, Napoli, Italy
e-mail: bianca.caiazzo@unina.it; angelo.coppola@unina.it; alberto.petrillo@unina.it; stefania.santini@unina.it

A. Masone et al. (eds.), *Optimization and Data Science: Trends and Applications*, AIRO Springer Series 6, https://doi.org/10.1007/978-3-030-86286-2_9

1 Introduction

The reduction of energy consumption is a major challenges in automotive field in order to enhance the environmental impact of mobility ecosystem [10]. Within this framework, the development of Connected and Autonomous Vehicles (CAVs) have the potential to enhance energy efficiency [18, 19, 23] by properly adapting their motion on the basis of surrounding traffic environment information (e.g. status of the road or of neighboring vehicles) obtained via Vehicle-to-Vehicle (V2V), Vehicle-to-Infrastructure (V2I) and/or Vehicle-to-Cloud (V2C) communications technologies [4–6, 21].

Over the past years, several strategies have been developed to achieve reduction in energy consumption by dealing with the factors that most affect it, i.e., the road slope and lead/predecessor vehicle behaviour. Authors in [22] proposed a distributed control framework aiming at reduce fuel consumption by operating all vehicles at a fuel-optimal speed profile, while in [1] an eco look-ahead controller predicting the optimum speed and acceleration profile on the basis of roadway slope for the lead vehicle of a platoon was developed. Due to its ability to take into account for different features at same time, MPC is widely used for optimization problems. For example, authors in [27] proposed a receding-horizon optimization based MPC taking into account for the previewed road slopes, while in [30] a Nonlinear MPC (NMPC) exploiting both terrain and preceding vehicle information was used. Furthermore, another crucial aspect in the dynamic behavior of platoon is related to inter-vehicle distance. Indeed, it is well known that smaller inter-vehicle gap can strongly reduce the aerodynamic drag, thus obtaining better performances in terms of energy consumption [2]. Common spacing policies, such as *Constant Spacing-Policy* (CS) and *Constant Time-Headway Policy* (CTH), require followers speed to converge the leader behavior in the time-domain. However, these spacing policies are typically used when constant references behaviour have to be track, while, in many practical situation, the reference speed signal could vary in spatial domain [3]. This consideration leads to the need of new energy-optimized spacing policies. Along this line, authors in [3] proposed a delay-dependent spacing policy allowing all vehicles within the platoon to track the same spatially varying reference speed profile. By considering also the road slope profile, authors in [26] compute the energy-optimal speed profile by exploiting Dynamic Programming and propose a variable spacing strategy based on the geometry relationship of each vehicle driving at optimal speed and its ideal speed under the CS policy.

This paper proposes an optimization procedure allowing to improve energy consumption of each vehicle within the platoon by acting on air-drag coefficient reduction, which vary as a function of inter-vehicle distance. Specifically, given optimal driving profile of the leader and a PI control strategy to ensure that followers track this desired energy-oriented behaviour, the proposed algorithm can improve energy performances by computing the optimal gap distances for each vehicles within the platoon w.r.t. its predecessor via Nonlinear Programming (NLP) Tool. Since the inter-vehicle distance should be as small possible in order to reduce aerodynamic drag, which decreases energy consumption, while a minimal spacing can increase the occurrence of collision among vehicles, safety constraints have been also embedded in the proposed strategy. Moreover, to avoid a continuous optimization procedure that might produce undesirable fast varying signal, the algorithm provides a suitable criterion to determine when the optimization procedure has to be run based on significant variation in driving and road slope profile.

Finally, the paper is organized as follows. In Sect. 2 we present vehicles, battery and consumption models, while the proposed optimization procedure is presented in Sect. 3. The effectiveness of the approach via numerical analysis is presented in Sect. 4, while conclusions are reported in Sect. 5.

2 Problem Statement

Consider N autonomous Electric Vehicles (EVs) plus a leader, indexed with 0, imposing the reference behaviour for the whole platoon. Vehicles are organized as a fleet and share their state information (e.g., position, speed and acceleration) with all the others vehicles through a V2V communication and proximity sensors [9].

2.1 *Autonomous Electric Vehicles Longitudinal Dynamics*

The longitudinal behavior of the i-th EV ($i = 1, 2, \cdots, N$) can be described by the following nonlinear dynamics [15, 17, 25]:

$$\begin{aligned}\dot{p}_i(t) &= v_i(t)\\ \dot{v}_i(t) &= \frac{\eta_i(t)}{R_i(t)m_i(t)}u_i(t) - gsin(\theta_i(t)) - gcos(\theta_i(t))\frac{C_r}{1000}(c_1 v(t) + c_2)\\ &\quad - \frac{\rho}{2m_i(t)}C_{D_i}(t)(1-\phi_i)A_{f_i}(t)v_i^2(t),\end{aligned} \tag{1}$$

where $p_i(t)$ $[m] \in \mathbb{R}$ and $v_i(t)$ [m/s] $\in \mathbb{R}$ are the position and the speed of the i-th vehicle, respectively; $u_i(t)$ [N m] is the control input that represents the vehicle propulsion torque, i.e. the driving/braking torque; $m_i(t)$ [kg] is the vehicle mass; $\eta_i(t)$ is the drive-train mechanical efficiency; $R_i(t)$ [m] is the wheel radius; C_r, c_1 and c_2 are the rolling resistance parameters that vary as a function of the road surface type, road condition, and vehicle tire type; ρ [kg/m^3] is the air density; $C_{D_i}(t)$ is the vehicle drag coefficient; $A_{f_i}(t)$ $[m^2]$ is the vehicle frontal area; ϕ_i is the air-drag reduction coefficient due to the platooning effects (according to [29]); g [m/s^2] is the gravity acceleration while $\theta_i(t)$ [rad] is the road-track slope.

Indicating with $x_i(t) = [p_i(t), v_i(t)]^\top \in \mathbb{R}^{2\times1}$ the i-th vehicle state vector, the nonlinear dynamic in (1) can be rewritten as follows:

$$\dot{x}_i(t) = \begin{bmatrix} v_i(t) \\ \varphi_i(v_i(t)) \end{bmatrix} + \begin{bmatrix} 0 \\ b_i \end{bmatrix} u_i(t) \tag{2}$$

where $b_i = \eta_i/(m_i R_i)$ while $\varphi_i(v_i(t)) \in \mathbb{R}$ is a continuously differentiable and bounded nonlinear vector field, defined as:

$$\begin{aligned}\varphi_i(v_i(t)) = & -gsin(\theta_i(t)) - gcos(\theta_i(t))\frac{C_r}{1000}(c_1 v_i(t) + c_2) \\ & - \frac{0.5}{m_i(t)}\rho C_{D_i}(t)(1-\phi_i)A_{f_i}(t)v_i^2(t).\end{aligned} \tag{3}$$

In the same way, the leader dynamics can be described by the following nonlinear system:

$$\dot{x}_i(t) = \begin{bmatrix} v_0(t) \\ \varphi_i(v_0(t)) \end{bmatrix} + \begin{bmatrix} 0 \\ b_0 \end{bmatrix} u_0(t) \tag{4}$$

where $x_0(t) = [p_0(t)\ v_0(t)]^\top$, being $p_0(t)$ [m] $\in \mathbb{R}$[m/s] and $v_0(t) \in \mathbb{R}$ the position and the velocity of the leading vehicle, respectively; $u_0(t)$ [N m] is the leader control input that represents the vehicle propulsion torque, i.e. the driving/braking torque, optimized via a MPC strategy as in [20]. Conversely, $u_i(t)$ in (1) is the control input for each follower selected according to the distributed PI control strategy proposed in [15], which updates its action based on the errors among the state information shared by the vehicles via the communication network as

$$\begin{aligned}u_i(t) = & - K_p \sum_{j=0}^{N} a_{ij}\Big(p_i(t) - p_j(t) - d_{ij}\Big) - K_i \sum_{j=0}^{N} a_{ij} \int_0^t \Big(p_i(s) - p_j(s) - d_{ij}\Big)ds \\ & - K_d \sum_{j=0}^{N} a_{ij}\Big(v_i(t) - v_j(t)\Big),\end{aligned} \tag{5}$$

where K_p, K_d and K_i are the proportional and integral control gains, respectively. In so doing, this distributed PI controller can guarantee that each follower within the platoon can track the leader energy-saving driving profile.

2.2 Battery Model

To model the EV battery pack of each vehicle i ($i = 1, \cdots, N$) within the platoon, according to [14], we consider an equivalent simplified electric circuit [14]. It consists of an internal voltage source $E_{oc,i}$, two ideal diodes, and two inner resistances $R^+_{in,i}$ and $R^-_{in,i}$ which represent the battery internal discharging and charging resistances, whose values depends on the actual value of the battery

State Of Charge (SOC) [14]. The voltage at the terminal of the battery is computed as:

$$V_{t,i} = E_{oc,i} - RI_{batt,i} = \begin{cases} E_{oc,i} - R^{+}_{in,i} I_{batt,i} & if discharging \\ E_{oc,i} - R^{-}_{in,i} I_{batt,i} & if charging \end{cases} \tag{6}$$

Indicating with $P_{req,i}$ the power required at the battery, the battery current $I_{batt,i}$ can be derived as:

$$I_{batt,i}(t) = \begin{cases} \dfrac{E_{oc,i} - \sqrt{E^2_{oc,i} - \frac{4R^{-}_{in,i} P_{req,i}}{n_{b,i}}}}{2R^{-}_{in,i}} & if discharging \\ \dfrac{E_{oc,i} - \sqrt{E^2_{oc,i} - \frac{4R^{-}_{in,i} P_{req,i}}{n_{b,i}}}}{2R^{+}_{in,i}} & if charging. \end{cases} \tag{7}$$

being $n_{b,i}$ the number of cells constituting the battery. The battery State of Charge can be, hence, computed as:

$$SOC_i(t) = \begin{cases} -\frac{1}{C_{batt,i}} \int_0^t I_{batt,i}(\tau) d\tau & if discharging \\ -\frac{\eta_{batt,i}}{C_{batt,i}} \int_0^t I_{batt,i}(\tau) d\tau & if charging. \end{cases} \tag{8}$$

where $C_{batt,i}$ is the battery capacity, while $\eta_{batt,i}$ is the recharging efficiency of the battery.

2.3 Power-Based Energy Consumption Estimation Model

Considering that the battery supplies a DC electric motor, for each vehicle i within the platoon ($i = 1, 2, \cdots, N$), the power-based EV Energy consumption Model is a quasi-steady backward model whose input are the instantaneous vehicle speed and the electric vehicle characteristics. The output of the model is the energy consumption (EC) [kW h/km] required by the vehicle for a specific drive cycle. To compute the EC, we first compute the power at electric motor according to the The Comprehensive Power-based EV Energy consumption Model (CPEM) [12]:

$$P_{em,i}(t) = \Big(m_i a_i - \phi_i(v_i(t))\Big) \frac{1}{\eta_i \eta_{em}} \tag{9}$$

where a [m/s^2] is the acceleration on the vehicle i; $\eta_{em} = 0.91$ is the electric motor efficiency, respectively.

The model takes into account the ability of EVs to recover energy during braking manoeuvres using a regenerative braking system. Hence, the effective electric power $P_{em_e ff,i}(t)$ is computed distinguishing between the traction and the regenerative braking mode:

$$P_{em_e ff,i}(t) = \begin{cases} P_{em,i}(t) & \text{if} P_{w,e}(t) \geq 0, \\ P_{em,i}(t) \cdot \eta_{rb,i}(t) & \text{if} P_{w,e}(t) < 0, \end{cases} \tag{10}$$

where $\eta_{rb,i}(t)$ is the regenerative braking efficiency computed according to [12].

Finally, the required electric power is obtained by considering also the auxiliary power loss $P_{aux} = 700[W]$:

$$P_{req,i}(t) = P_{em_eff,i}(t) + P_{aux}. \tag{11}$$

Given the required power and the distance d_i $[km]$ travelled by vehicle i, the energy consumption $[\frac{\text{kW h}}{\text{km}}]$ can be computed according to [12]:

$$EC_i = \frac{1}{3600000} \int_0^t P_{req,i}(\tau)\, d\tau \; \frac{1}{d_i}. \tag{12}$$

3 Optimization Procedure

Let $v_i^\star(t)$ and $a_i^\star(t)$ the optimal speed and acceleration profile of each vehicle, properly computed according to MPC together with distributed-PI procedures. Let $d_{i,i-1}^\star(t)$ the optimal time-varying inter-vehicle distance between vehicle i and its predecessor $i-1$ to be properly computed for energy saving purposes. To deal with the problem of inter-vehicle distance optimization, we exploit the distance-dependent air drag coefficient expressed as [24]

$$C_{D_i}(d_{i,i-1}^\star) = C_{D_i,0}\left(1 - \frac{C_{D_i,1}}{C_{D_i,2} + d_{i,i-1}^\star}\right) \tag{13}$$

being $C_{D_i,0}$ the i-th vehicle drag coefficient in the absence of any slipstream (i.e. the drag coefficient of the lead vehicle), while $C_{D_i,1}$ and $C_{D_i,2}$ positive constants obtained by regressing the experimental data presented in [13].

By leveraging the above distance-dependent air drag coefficient (13), the required power by i-th vehicle in order to achieve tracking performance can be, hence, expressed as the following distance-dependent nonlinear function:

$$\begin{aligned} P_i(d_{i,i-1}^\star(t)) = &\Big(m a_i^\star(t) + mg\sin(\theta_i(t)) + mg\cos\theta_i(t)\frac{C_r}{1000}(c_1 v_i^\star(t) + c_2) \\ &+ \frac{\rho}{2} A_{f_i} C_{D_i}(d_{i,i-1}^\star)(v_i^\star)^2(t)\Big) v_i^\star(t), \end{aligned} \tag{14}$$

where $\theta_i(t)$ is the road slope profile, which can be accurately obtained by combining the GPS and geographic information system [8, 11, 26]. In order to achieve a further energy consumption reduction while considering safety constraints, for each vehicle within the platoon the following optimization problem is given:

$$\begin{aligned} &\min_{d_{i,i-1}} P_i(d_{i,i-1}(t)) \\ &s.t. \\ &d_{i,i-1}^{min}(t) \le d_{i,i-1}^\star(t) \le d_{i,i-1}^{max}(t), \end{aligned} \tag{15}$$

where $P_i(d_{i,i-1}(t))$ is the required power as in (14), while $d^{min}_{i,i-1}(t)$ and $d^{max}_{i,i-1}(t)$ refer to the minimum and maximum allowed inter-vehicle distances between vehicles i and $i-1$ in order to ensure emergency braking maneuvers as well as air-drag reduction. Specifically, these distances constraints can be calculated, for each time instant, using the lower and upper bound of the vehicle time-headway, i.e. h_{min} and h_{max} respectively, as

$$d^{min}_{i,i-1} = d_{st} + h_{min} v_i^\star(t), \qquad d^{max}_{i,i-1} = d_{st} + h_{max} v_i^\star(t), \tag{16}$$

According to the technical literature [28], the typical value for lower and upper time-headway are given as $h_{min} = 0.4$[s] and $h_{max} = 1$[s]. The aim is to avoid a continuous optimization procedure due to instability problem that might arises due to the presence of fast varying signal. Moreover, since the optimal inter-vehicles gap is strongly affected by variations in driving and road slope profiles [7], we propose an inter-vehicle optimization procedure as in Fig. 1. More specifically, by introducing road slope profile derivative $\dot{\theta}_i(t)$ and disagreement vectors $ds = a_i^\star(t) - a_i^\star(t-1)$ and $dt = \dot{\theta}_i(t) - \dot{\theta}_i(t-1)$ to capture significant changes in driving

Algorithm 1 Computing the Optimal Desired Inter-vehicle Gap for the Vehicle i.

Declarations:
Let $v_i^\star$ and $a_i^\star$ be the i-vehicle speed and acceleration computed via PI controller (5).
Let θ_i and $\dot{\theta}_i$ be the road slope angle accurately obtained and its derivative.
Let h_{min} and h_{max} the lower and upper limits for time headway properly fixed.
Let $k = length(v_i) = length(a_i)$ the total number of iterations.
Let t_1 and t_2 the pre-fixed threshold for tacking in to account acceleration and road slope angle variations.
Create vectors ds and dr that capture significant variation in acceleration magnitude, i.e. a change between $a(i)$ and $a(i-1)$, and a change in slope road derivative profile, i.e. between $\dot{\theta}(i)$ and $\dot{\theta}(i-1)$ respectively.
Initialization:
Initialize the null vector $ds = [\,] \in \mathbb{R}^k$
Initialize the null vector $dr = [\,] \in \mathbb{R}^k$
for $i = 1$ **to** k **do**
  $ds(i-1) = a(i) - a(i-1)$;
  $dt(i-1) = \dot{\theta}(i) - \dot{\theta}(i-1)$;
end
Initialize a null vector $d^\star_{i,i-1} = [\,] \in \mathbb{R}^k$.
Optimize the intervehicular distance every time when the threshold t_1 and t_2 are exceeded in ds or dr respectively.
for $j = 1$ **to** k **do**
  if $abs(ds(j)) > t_1 \,\|\, abs(dr(j)) > t_2$ **then**
    $d^{min}_{i,i-1}(j) = d_{st} + h_{min} v_i^\star(j)$;
    $d^{max}_{i,i-1}(j) = d_{st} + h_{max} v_i^\star(j)$;
    $A = [-1; 1]$;
    $b = [d^{min}_{i,i-1}(j); d^{max}_{i,i-1}(j)]$;
    Solve problem (16) and update $d^\star_{i,i-1}$;
    $d^\star_{i,i-1}(j) = x(j)$;
  else
    $d^\star(j) = d^\star(j-1)$;
end

Fig. 1 Optimization algorithm

and road slope profiles, the proposed Algorithm allows computing the optimal inter-vehicles distance vector $d^{\star}_{i,i-1}(t)$ for each vehicle $i = 1, \dots, N$, based on properly selected threshold. In so doing, the proposed Algorithm provides a suitable criterion to determine the most influential time-instant when the optimization procedure has to be run. In this paper, the optimization problem as in (15) is solved through a Nonlinear Programming (NLP) algorithm via Matlab Optimization Toolbox.

4 Numerical Results

Here, we prove the effectiveness of the proposed optimization procedure in improve the energy performance of a platoon of electric vehicles. More specifically, we consider a platoon consisting of 5 electric vehicles plus a leader sharing information via the *leader-predecessor-follower* (L-P-F) communication topology [31] and moving along a road highway segment whose altitude profile is reported in Fig.3a. Furthermore, the leader is equipped with MPC controller [20], while followers can track the optimized reference leader behaviour via the distributed PI control strategy in (5). To disclose the benefit of the proposed optimization algorithm, we consider two simulation scenario: (i) *base* scenario, where the inter-vehicle distances are computed via CTH spacing policy, i.e. $d_{i,i-1} = d^{st} + h * v_i(t)$, with $h = 0.8$[s] [16]); (ii) *optimized* scenario, where Fig. 1 is exploited. The aim is to evaluate how the optimal time-varying spacing policy, computed according to Fig. 1, can further reduce the energy consumption w.r.t. the case where no optimization procedure is considered. The numerical analysis is carried out via the Matlab/Simulink simulation platform, while vehicles parameters are listed in Table 1.

Results in Fig. 2a,b show the behavior of the platoon in *base* scenario, in terms of speed and acceleration profiles respectively, while the inter-vehicle distance of each vehicle within the platoon w.r.t. its predecessor is represented in Fig. 2c. For the resulting driving profile in Fig. 2 and for road slope profile as in Fig.3, the optimization Fig. 1 provides the optimal time-varying inter-vehicle spacing profile as in (4)c. Note that, within the optimization procedure, for each vehicle we consider

Table 1 Heterogeneous nonlinear vehicles parameters

Vehicle ID	m_i [kg]	η_i [−]	R_i [m]	$C_{D,i}$ [−]	$A_{f,i}$ [m^2]	a_{max} [m/s^2]	a_{min} [m/s^2]	$C_{batt,i}$ [Ah]	$n_{b,i}$ [−]	$\eta_{batt,i}$ [−]
0	1545	0.89	0.3060	0.28	2.3315	2.5	−6.0	65	96	0.97
1	1015	0.89	0.2830	0.30	2.1900	2.5	−6.0	65	96	0.97
2	1375	0.89	0.2880	0.24	2.4000	2.5	−6.0	65	96	0.97
3	1430	0.89	0.3284	0.28	2.4600	2.5	−6.0	65	96	0.97
4	1067	0.89	0.2653	0.29	2.1400	2.5	−6.0	65	96	0.97
5	1155	0.89	0.2880	0.33	2.0400	2.5	−6.0	65	96	0.97

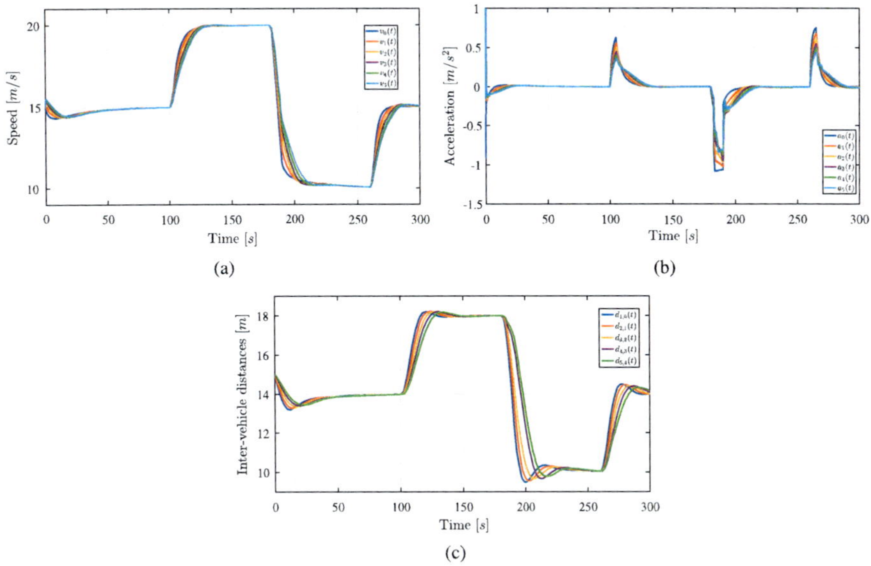

Fig. 2 Results of *base* scenario. Time history of: *a*) vehicles speed $v_i(t),\ i = 0, \ldots, 5$; *b*) vehicles acceleration $a_i(t),\ i = 0, \ldots, 5$; *c*) inter-vehicle distances $d_{i,i-1}(t),\ i = 1, \ldots, 5$

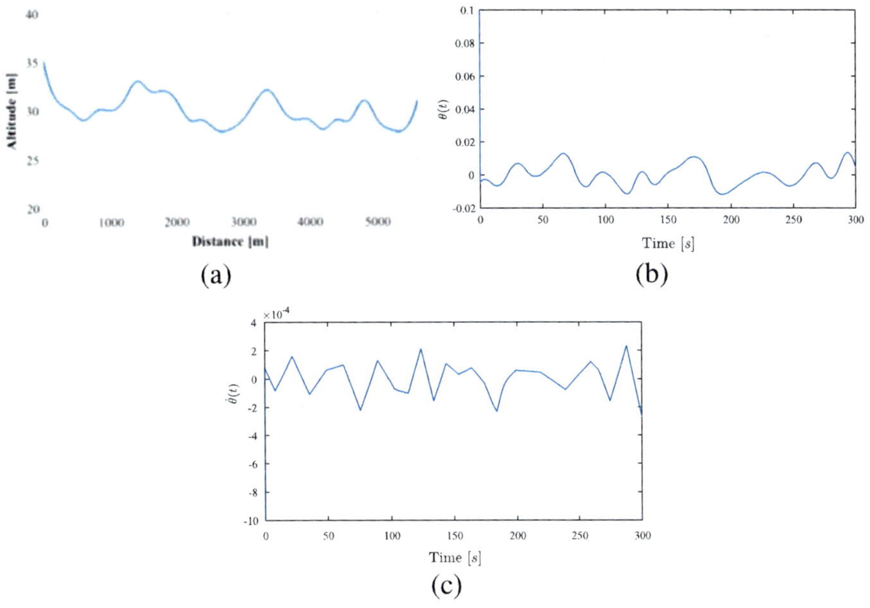

Fig. 3 Non-flat road features: (**a**) altitude profile; (**b**) road slope profile $\theta(t)$; (**c**) road slope variation $\dot{\theta}(t)$

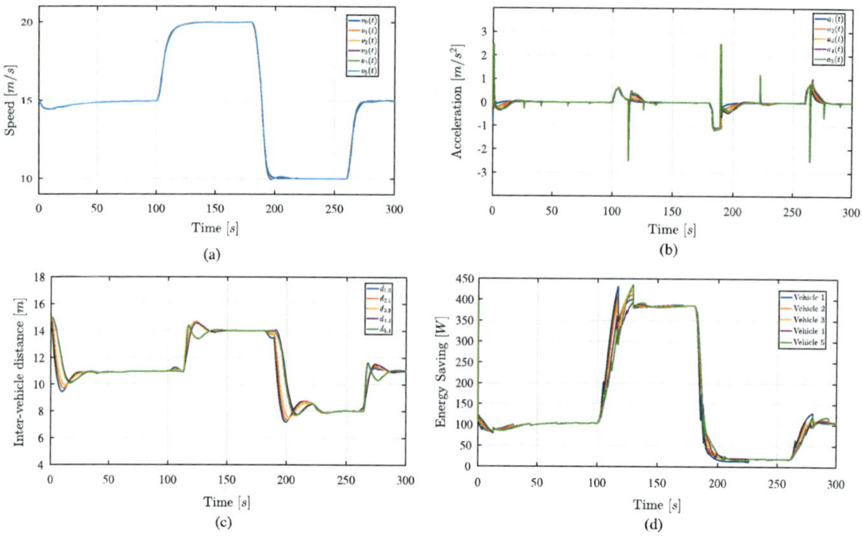

Fig. 4 Results of *optimized* scenario. Time history of: (**a**) vehicles speed $v_i(t)$, $i = 0, \dots, 5$; (**b**) vehicles acceleration $a_i(t)$, $i = 0, \dots, 5$; (**c**) inter-vehicle distances $d_{i,i-1}(t)$, $i = 1, \dots, 5$; *d*) Energy Saving w.r.t. *base* scenario

the slope profile of the road stretch travelled by the vehicle itself. Here, for the sake of brevity, we report just an exemplary one as in Fig 3. Hence, by applying the suggested optimization procedure, we obtain results in Fig. 4. Specifically, Fig. 4a,b disclose the time-history of vehicles speed and acceleration profiles, respectively, while Fig. 4c highlights the reduction of the gap distance w.r.t. Fig. 2c. To prove the benefit of the proposed optimization technique in terms of energy saving, we also compare the power required by a vehicle for tracking the reference driving profile imposed by the leader according to (11) both in the *base* scenario and in *optimized* scenario. The difference between the power computed in both cases are shown in Fig.4d, which corroborates the effectiveness of the proposed strategy in improve the energy saving. Finally, to quantify the energy reduction for each vehicle w.r.t *base* scenario, we compute the percentage variation of energy consumption. Results, reported in Table 2, confirm a average energy reduction of 2.7% for the entire platoon.

Table 2 Reduction of the energy consumption [kWh/km] in percentage for each vehicle *i*

Configuration	Vehicle 1	Vehicle 2	Vehicle 3	Vehicle 4	Vehicle 5	Mean
Time-varying spacing policy	−2.7129	−2.7038	−2.7090	−2.6804	−2.6944	−2.7000

5 Conclusion

In this paper, an integrated optimization procedure for the optimization of inter-vehicles distances is investigated for an autonomous platoon of heterogeneous electric vehicles. The proposed optimization approach allows to improve energy consumption of each vehicle by acting on air-drag coefficient, which vary as a function of inter-vehicle distance. Specifically, the proposed Algorithm, on the basis of both driving and road slope profiles, computes the optimal time-varying gap distances for each vehicle within the platoon w.r.t. its predecessor via NLP Tool. Safety constraints have been also embedded in the proposed strategy to guarantee safety driving conditions, i.e., avoid collision among vehicles. Finally, numerical analysis have been carried out to evaluate the effectiveness of the proposed approach in guaranteeing energy saving with respect to a base scenario where the typical CTH spacing policy has been employed. Results highlight how the proposed algorithm ensure an improvement of energy performance for each vehicle within the platoon w.r.t. the base scenario, resulting in an average reduction of about 2.7% for the whole platoon.

References

1. Ahn, K., Rakha, H.A., Park, S.: Eco look-ahead control of battery electric vehicles and roadway grade effects. Transp. Res. Rec. **2674**(10), 429–437 (2020)
2. Alam, A., Besselink, B., Turri, V., Mårtensson, J., Johansson, K.H.: Heavy-duty vehicle platooning for sustainable freight transportation: A cooperative method to enhance safety and efficiency. IEEE Control Syst. Mag. **35**(6), 34–56 (2015)
3. Besselink, B., Johansson, K.H.: String stability and a delay-based spacing policy for vehicle platoons subject to disturbances. IEEE Trans. Autom. Control **62**(9), 4376–4391 (2017)
4. Bifulco, G.N., Caiazzo, B., Coppola, A., Santini, S.: Intersection crossing in mixed traffic flow environment leveraging v2x information. In: 2019 IEEE International Conference on Connected Vehicles and Expo (ICCVE), pp. 1–6. IEEE (2019)
5. Boccia, M., Masone, A., Sforza, A., Sterle, C.: A column-and-row generation approach for the flying sidekick travelling salesman problem. Transport. Res. C Emerg. Technol. **124**, 102913 (2021)
6. Boccia, M., Masone, A., Sforza, A., Sterle, C.: An exact approach for a variant of the fs-tsp. Transport. Res. Procedia **52**, 51–58 (2021)
7. Caldas, K.A.Q., Grassi, V.: Eco-cruise nmpc control for autonomous vehicles. In: 2019 19th International Conference on Advanced Robotics (ICAR), pp. 356–361. IEEE (2019)
8. Castiglione, L.M., Falcone, P., Petrillo, A., Romano, S.P.: and Stefania Santini. Cooperative intersection crossing over 5g. IEEE/ACM Trans. Netw. (2020)
9. Coelingh, E., Solyom, S.: All aboard the robotic road train. IEEE Spectrum **49**(11), 34–39 (2012)
10. Di Costanzo, L., Coppola, A., Pariota, L., Petrillo, A., Santini, S., Bifulco, G.N.: Variable speed limits system: A simulation-based case study in the city of naples. In: 2020 IEEE International Conference on Environment and Electrical Engineering and 2020 IEEE Industrial and Commercial Power Systems Europe (EEEIC/I&CPS Europe), pp. 1–6. IEEE (2020)

11. Di Vaio, M., Falcone, P., Hult, R., Petrillo, A., Salvi, A., Santini, S.: Design and experimental validation of a distributed interaction protocol for connected autonomous vehicles at a road intersection. IEEE Trans. Veh. Technol. **68**(10), 9451–9465 (2019)
12. Fiori, C., Ahn, K., Rakha, H.A.: Power-based electric vehicle energy consumption model: Model development and validation. Applied Energy **168**, 257–268 (2016)
13. Hucho, W., Sovran, G.: Aerodynamics of road vehicles. Annu. Rev. Fluid Mech. **25**(1), 485–537 (1993)
14. Maia, R., Silva, M., Araújo, R., Nunes, U.: Electrical vehicle modeling: A fuzzy logic model for regenerative braking. Expert Syst. Appl. **42**(22), 8504–8519 (2015)
15. Manfredi, S., Petrillo, A., Santini, S.: Distributed pi control for heterogeneous nonlinear platoon of autonomous connected vehicles. IFAC-PapersOnLine **53**(2), 15229–15234 (2020)
16. Nowakowski, C., Shladover, S.E., Cody, D., Bu, F., O'Connell, J., Spring, J., Dickey, S., Nelson, D.: Cooperative adaptive cruise control: Testing drivers' choices of following distances. Technical report (2011)
17. Pariota, L., Coppola, A., Di Costanzo, L., Di Vico, A., Andolfi, A., D'Aniello, C., Bifulco, G.N.: Integrating tools for an effective testing of connected and automated vehicles technologies. IET Intell. Transp. Syst. **14**(9), 1025–1033 (2020)
18. Pariota, L., Di Costanzo, L., Coppola, A., D'Aniello, C., Bifulco, G.N.: Green light optimal speed advisory: a c-its to improve mobility and pollution. In: 2019 IEEE International Conference on Environment and Electrical Engineering and 2019 IEEE Industrial and Commercial Power Systems Europe (EEEIC/I&CPS Europe), pp. 1–6. IEEE (2019)
19. Petrillo, A., Pescapé, A., Santini, S.: A secure adaptive control for cooperative driving of autonomous connected vehicles in the presence of heterogeneous communication delays and cyberattacks. IEEE Trans. Cybern. **51**(3), 1134–1149 (2021)
20. Sajadi-Alamdari, S.A., Voos, H., Darouach, M.: Nonlinear model predictive extended eco-cruise control for battery electric vehicles. In: 2016 24th Mediterranean Conference on Control and Automation (MED), pp. 467–472. IEEE (2016)
21. Sarker, A., Shen, H., Rahman, M., Chowdhury, M., Dey, K., Li, F., Wang, Y., Narman, H.S.: A review of sensing and communication, human factors, and controller aspects for information-aware connected and automated vehicles. IEEE Trans. Intell. Transport. Syst. **21**(1), 7–29 (2019)
22. Turri, V., Besselink, B., Johansson, K.H.: Cooperative look-ahead control for fuel-efficient and safe heavy-duty vehicle platooning. IEEE Trans. Control Syst. Technol. **25**(1), 12–28 (2016)
23. Vahidi, A., Sciarretta, A.: Energy saving potentials of connected and automated vehicles. Transport. Res. C Emerg. Technol. **95**, 822–843 (2018)
24. Wasserburger, A., Schirrer, A., Hametner, C.: Stochastic optimization for energy-efficient cooperative platooning. In: 2019 IEEE Vehicle Power and Propulsion Conference (VPPC), pp. 1–6. IEEE (2019)
25. Wu, Y., Li, S.E., Cortés, J., Poolla, K.: Distributed sliding mode control for nonlinear heterogeneous platoon systems with positive definite topologies. IEEE Trans. Control Syst. Technol. (2019)
26. Xu, L., Zhuang, W., Yin, G., Bian, C.: Energy-oriented cruising strategy design of vehicle platoon considering communication delay and disturbance. Transport. Res. C Emerg. Technol. **107**, 34–53 (2019)
27. Xu, S., Peng, H.: Design and comparison of fuel-saving speed planning algorithms for automated vehicles. IEEE Access **6**, 9070–9080 (2018)
28. Yanakiev, D., Kanellakopoulos, I.: Nonlinear spacing policies for automated heavy-duty vehicles. IEEE Trans. Veh. Technol. **47**(4), 1365–1377 (1998)
29. Zabat, M., Stabile, N., Farascaroli, S., Browand, F.: The aerodynamic performance of platoons: A final report (1995)

30. Zhang, S., Luo, Y., Wang, J., Wang, X., Li, K.: Predictive energy management strategy for fully electric vehicles based on preceding vehicle movement. IEEE Trans. Intell. Transport. Syst. **18**(11), 3049–3060 (2017)
31. Zheng, Y., Li, S.E., Wang, J., Cao, D., Li, K.: Stability and scalability of homogeneous vehicular platoon: Study on the influence of information flow topologies. IEEE Trans. Intell. Transport. Syst. **17**(1), 14–26 (2015)

Optimization-Based Assessment of Initial-State Opacity in Petri Nets

Gianmaria De Tommasi, Carlo Motta, Alberto Petrillo, and Stefania Santini

Abstract When dealing with security and safety problems, Discrete Events Systems (DESs) could be a convenient way to model the behavior of distributed dynamical systems. Among the different DES mathematical tools, Petri Nets (PNs), by benefiting from a twofold representation, i.e. a graphical and a mathematical one, can be exploited for effectively tackling some security problems in the DES context such as the opacity one. This latter property is related to the capability of hiding a secret to external observers. When the secret is modeled by the initial marking (state) of a PN, the problem is known in literature as Initial-State Opacity (ISO). A DES is said to be ISO if, for every trajectory originating from a secret state, there exists another trajectory originated from a non-secret state, such that both of them are equivalent from an external observer (potentially malicious) point of view. Therefore, in an opaque system, such intruder can never determine whether the system started from a secret state or from a non-secret one. In this paper, leveraging the mathematical representation of PNs, we present a sufficient condition which permit to assess if a system is not opaque, by solving a feasibility problem with integer optimization variables. Specifically, the proposed approach starts from the ISO definition and, then, characterizes the aforementioned non-opacity condition as a set of linear constraints that, if not satisfied, imply the system to be not ISO.

Keywords Opacity · Petri nets · ILP problems

G. De Tommasi · C. Motta (✉) · A. Petrillo · S. Santini
Dipartimento di Ingegneria Elettrica e delle Tecnologie dell'Informazione, Università degli Studi di Napoli Federico II, Napoli, Italy
e-mail: detommas@unina.it; carlo.motta@unina.it; alberto.petrillo@unina.it; stefania.santini@unina.it

A. Masone et al. (eds.), *Optimization and Data Science: Trends and Applications*, AIRO Springer Series 6, https://doi.org/10.1007/978-3-030-86286-2_10

1 Introduction

Modern control systems are becoming more open to the cyberworld and, as such, are more vulnerable to cyber-attacks [2, 9]. As a consequence, a major challenge is the design of supervisory and control systems that are resilient to them [6, 13, 15]. Indeed, in a distributed system, information leaks and deceptions represent a threat to the privacy and security of the system itself, since they may enable external cyber attackers to infer information about the system state, and consequently interact in a malicious way with safety-critical functions. This can be accomplished in two different ways, denoted *active* or *passive* attacks. In active attacks, the attacker's goal is to inflict damage on the system, while in passive attacks the attacker's goal is to learn secrets about the system [14]. In the first case a security problem arises, while the latter deals with system privacy.

Security and privacy problems can be effectively modeled in the framework of Discrete Event Systems (DESs, [10]). Two main information-flow concepts have been successfully used to characterize privacy, and hence passive attacks, when the CPSs are modelled at the DES level: opacity [5, 12, 16, 17] and non-interference [6, 7]. When dealing with opacity, the secret is either a sublanguage of the language generated by the plant model (language-based opacity), or a system state, either initial, current or final (state-based opacity). In an opaque system, a user with full knowledge of the model, but with partial capabilities about the observation of the event occurrences cannot infer any secret, no matter for how long the system dynamic is partially observed. A system can be designed as such, otherwise supervisory control can be used to enforce opacity by restricting the closed-loop behaviour in presence of controllable events [15]. An alternative approach consists in designing *ad hoc* obfuscation/insertion functions to fool the malicious observer [13].

Here we deal with Initial State Opaque (ISO) in DESs. Formally, a DES is said to be ISO if, for every trajectory originating from a secret state, there exists another trajectory originated from a non-secret state, such that both of them are equivalent from an external, potentially malicious, observer's point of view. Therefore, in an opaque system, such intruder can never determine whether the system started from a secret state or from a non-secret one.

In this paper we introduce a sufficient condition to conclude if a DES modeled as a PN system is not ISO. Such a sufficient condition is based on the solution of Integer Linear Programming (ILP) optimization problems. Optimization-based approaches does not require the explicit computation of the PN reachability set, hence they are particularly well suited for models with a high level of parallelism (for a quantitative comparison between optimization-based and graph-based approaches the interested readers can refer to [3]). Although a similar approach has been proposed to deal with both non-interference [6] and language-based opacity [5], to the best of the authors' knowledge the result presented in this paper represent a first attempt to address the state opacity problem by means of an optimization-based approach.

The paper is structured as follows: in the next section the notation adopted for PNs is introduced, together with the definition of the considered opacity properties. Section 3 gives the main result of this paper, namely a sufficient condition to verify if a PN system is not ISO, based on the solution of ILP problems. Moving forward, Sect. 4 some examples are presented in order to show the effectiveness of the proposed approach, while conclusive remarks and possible future research directions are finally discussed in Sect. 5.

2 Backgrounds

As already mentioned in the previous section, in this work we deal with opacity in DESs modeled as Petri Net (PN) systems. Therefore, in this background section, we first introduce the notation adopted to describe PNs. Afterwards the concept of ISO is formally given. The interested reader can find more details on PNs in [8], while a comprehensive discussion about opacity in DES can be found in [17].

2.1 *Basic Petri Nets notation*

PNs are *Place/Transition* (P/T) net defined as 4-tuple $N = (P, T, \mathbf{Pre}, \mathbf{Post})$, where P is the set of m places (represented by circles), T is the set of n transitions (represented by boxes), and $\mathbf{Pre} : P \times T \mapsto \mathbb{N}$ ($\mathbf{Post} : P \times T \mapsto \mathbb{N}$) is the pre- (post-) incidence matrix. $\mathbf{Pre}(p, t) = \omega$ ($\mathbf{Post}(p, t) = \omega$) means that there is an arc with weight ω from p to t (from t to p); $\mathbf{C} = \mathbf{Post} - \mathbf{Pre}$ is the incidence matrix.

Given a P/T net it is possible to assign a non-negative number of tokens to each of its places by means of the *marking* function $\vec{m} : P \mapsto \mathbb{N}$; graphically the tokens are represented as black dots, see the example shown in Fig. 1. The marking of a PN represents the state of the system and is usually specified as a vector $\vec{m} \in \mathbb{N}^m$, being $\vec{m}_0$ the initial marking. A PN system $\mathscr{S} = \langle N, \vec{m}_0 \rangle$ is given by the net and its initial marking; as an example the initial marking for the net system shown in Fig. 1 is equal to $\vec{m}_0 = (2\ 0\ 0)^T$.

The marking in a PN system, i.e. its state, evolves following the *firing* of the net transitions. In particular, a transition t is enabled under the marking $\vec{m}$ if and only if $\vec{m} \geq \mathbf{Pre}(\cdot, t)$; the transition enabling is denoted as $\vec{m}[t\rangle$. An enabled transition t may fire, yielding the marking $\vec{m}' = \vec{m} + \boldsymbol{C}(\cdot, t)$, and this is denoted as $\vec{m}[t\rangle\vec{m}'$, while $\vec{m}\neg[t\rangle$ denotes that the transition t is not enabled under $\vec{m}$. A *firing sequence* from $\vec{m}$ is a sequence of transitions $\sigma = t^1 t^2 \ldots t^k$ such that $\vec{m}[t^1\rangle\vec{m}_1[t^2\rangle\vec{m}_2 \ldots [t^k\rangle\vec{m}_k$ and this is denoted as $\vec{m}[\sigma\rangle\vec{m}_k$. The notations $\vec{m}[\sigma\rangle$ and $\vec{m}\neg[\sigma\rangle$ denote an enabled and a disabled sequence under the marking $\vec{m}$, respectively. The length of a sequence σ is denoted with $|\sigma|$, and the symbol ε is used to denote the *empty* sequence or the *silent* transition; it

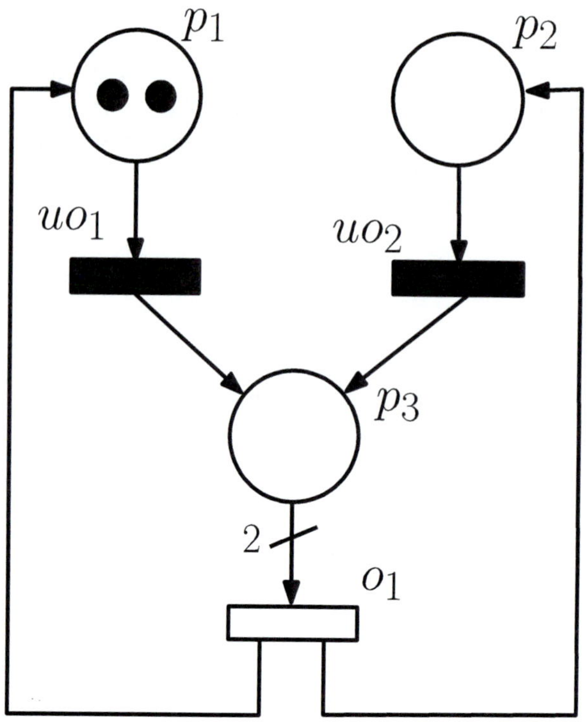

Fig. 1 An example of Petri net system

is $\sigma\varepsilon = \varepsilon\sigma = \sigma$, and $|\varepsilon| = 0$; moreover it is $\vec{m}[\varepsilon\rangle\vec{m}$. A marking $\vec{m}'$ is said to be reachable from $\vec{m}_0$ *iff* there exists a sequence σ such that $\vec{m}_0[\sigma\rangle\vec{m}$. $R(N, \vec{m}_0)$ denotes the set of reachable markings of the Petri net system $\mathscr{S}$. The language of a Petri net system $\mathscr{S}$ is defined as follows[1]

$$L(N, \vec{m}_0) = \{\sigma \in T^* \mid \vec{m}_0[\sigma\rangle\} .$$

It is possible to associate a vector to any a firing sequence σ by means of the function $\vec{\sigma} : T \mapsto \mathbb{N}$, where $\vec{\sigma}(t)$ represents the number of occurrences of t in σ. The vector $\vec{\sigma}$ is called the *firing count vector* of σ. The notation $\vec{\sigma} = \pi(\sigma)$ is used to denote that $\vec{\sigma}$ is the firing count vector of σ.

[1] The notation T^* denotes the Kleene closure of T (see [10, Ch. 2]).

If $\vec{m}_0[\sigma\rangle\vec{m}$, then it is possible to write the vector equation

$$\vec{m} = \vec{m}_0 + \boldsymbol{C} \cdot \vec{\sigma}\,, \tag{1}$$

which is called the *state equation* of the net system.

The next result, taken from [11], gives a necessary and sufficient condition that must be fulfilled by every sequence with finite length enabled under the marking $\vec{m}$.

Lemma 1 ([11]) *There exists ρ integer vectors $\vec{s}_1, \ldots, \vec{s}_\rho \in \mathbb{N}^n$ with $\rho \leq |\sigma|$ such that the following linear constraints are* fulfilled

$$\begin{aligned}
&\vec{m}_0 \geq \mathbf{Pre} \cdot \vec{s}_1 \\
&\vec{m}_0 + \boldsymbol{C} \cdot \vec{s}_1 \geq \mathbf{Pre} \cdot \vec{s}_2 \\
&\ldots \\
&\vec{m}_0 + \boldsymbol{C} \cdot \sum_{i=1}^{\rho-1} \vec{s}_i \geq \mathbf{Pre} \cdot \vec{s}_\rho
\end{aligned} \tag{2a}$$

$$\sum_{i=1}^{\rho} \vec{s}_i = \pi(\sigma) \tag{2b}$$

iff there exists at least one sequence σ, which is enabled under the marking $\vec{m}_0$ such that $\pi(\sigma) = \vec{\sigma}$.

Remark 1 In this paper we deal with bounded net systems, i.e. systems whose cardinality of $R(N, \vec{m}_0)$ is finite. For this class of systems, there exists an integer $\mathscr{I}_{\min}$ such that, if $\rho \geq \mathscr{I}_{\min}$, for each $\vec{m} \in R(N, \vec{m}_0)$ there exists at least one set of vectors $\vec{s}_1, \ldots, \vec{s}_\rho$ that fulfill the constraints (2a) and

$$\vec{m} = \vec{m}_0 + \boldsymbol{C} \cdot \sum_{i=1}^{\rho} \vec{s}_i\,.$$

In other words, for bounded net systems $\mathscr{I}_{\min}$ integer vectors are sufficient to describe the reachability set $R(N, \vec{m}_0)$ by means of the constraints (2a). For a bounded but *non live* net system, an estimation of $\mathscr{I}_{\min}$ can be carried out by using the reachability graph of the system itself. For bounded and live systems and estimation of $\mathscr{I}_{\min}$ can be computed exploiting the concept of T-invariant (for a more comprehensive discussion on this issue, the interested reader can refer to [4, Sec. 3]). ▲

In the next section we will deal with opacity in DESs modeled with PN systems. In this context, external users can only partially observe the system behaviour. In order to model this partial observability, the set T is partitioned into the two disjoint sets of observable (represented by empty boxes) and unobservable transitions (represented by filled boxes), named respectively T_o and T_{uo}. Given a sequence $\sigma \in T^*$, its observation is the output of the natural projection function $Pr : T^* \mapsto T_o^*$, which is recursively defined as $Pr(\sigma t) = Pr(\sigma)Pr(t)$, with $\sigma \in T^*$ and $t \in T$; moreover, $Pr(\sigma) = t$ if $t \in T_o$, while $Pr(t) = \epsilon$ if $t \in T_{uo}$.

Remark 2 In order to ease the notation, in Sect. 3 we will refer to $\mathbf{Pre}_o$ ($\mathbf{Pre}_{uo}$) as a partition of the pre-incidence matrix obtained by excluding the columns associated to the non-observable (observable) transitions. The same goes for $\mathbf{Post}_o$ ($\mathbf{Post}_{uo}$) and $\boldsymbol{C}_o$ ($\boldsymbol{C}_{uo}$) ▲

2.2 *Initial State Opacity in Petri Nets*

In the next section we will give a sufficient condition to check if a PN system is not initial state opaque. When dealing with ISO in PNs the information we are interested in hiding is the initial marking of the net, although the intruder knows the structure of the system—i.e., the net topology is known—he/she does not have a precise information about the initial state. For this reason, we will consider net systems where there is an uncertainty on the initial marking. To this aim we define $\mathcal{M}_0 \subseteq \mathbb{N}^m$ as the set of all the initial markings, that is we assume that $\vec{m}_0$ belongs to any of the markings in the set $\mathcal{M}_0$. Furthermore, $\mathcal{M}_0$ is split into two disjoint subsets: the set of secret markings $\mathcal{M}_s \subset \mathcal{M}_0$, and the set of non-secret markings $\mathcal{M}_{ns} \subset \mathcal{M}_0$, being $\mathcal{M}_s \cap \mathcal{M}_{ns} = \emptyset$. When dealing with systems with uncertain initial marking, we adopt the notation $\mathcal{S} = \langle N , \mathcal{M}_0 \rangle$, implying that $\vec{m}_0 \in \mathcal{M}_0$.

Given the two subsets $\mathcal{M}_s$ and $\mathcal{M}_{ns}$, we can retrieve the definition of ISO system by extending as follows the one given in [17] to the case of PN systems.

Definition 1 A net system $\mathcal{S}$ with uncertain initial marking belonging to $\mathcal{M}_0$ is ISO *if and only if*

$$\forall\, \vec{m}_s \in \mathcal{M}_s \text{ and } \forall\, \sigma \in L(N, \vec{m}_s)\,, \exists\, \vec{m}_{ns} \in \mathcal{M}_{ns} \text{ and } \exists\, \sigma' \in L(N, \vec{m}_{ns}) \text{ s.t.}$$
$$Pr(\sigma) = Pr(\sigma')\,. \qquad (3)$$

■

3 Main Results

This section gives the main contribution of this paper, namely a sufficient condition to check if a net system is not ISO. Therefore, before presenting such a main result, let us first derive the definition of a not ISO system. From Definition 1, the following readily follows.

Definition 2 A net system $\mathscr{S}$ with uncertain initial state belonging to $\mathscr{M}_0$ is not ISO if and only if

$$\exists\ \vec{m}_s \in \mathscr{M}_s \text{ and } \exists\ \sigma \in L(N, \vec{m}_s)\ ,\ \forall\ \vec{m}_{ns} \in \mathscr{M}_{ns} \text{ and } \forall\ \sigma' \in L(N, \vec{m}_{ns}) \text{ s.t.}$$
$$Pr(\sigma) \neq Pr(\sigma')\ . \qquad (4)$$

■

In other words, from Definition 2, a PN system with multiple secret and non-secret initial markings is not ISO *iff* at least one secret marking has a firing sequence whose projection on the observable transitions is different from the projection of every firing sequence generated by the set of not secret initial markings.

We can now state the following theorem, that exploits the mathematical representation of PN systems to recast Definition 2 in terms of a set of feasibility problems with integer optimization variables. It should be noticed that the proposed result represents only a *sufficient condition*. Indeed, if at last one of the considered feasibility problems does not admit a solution, then the system $\mathscr{S}$ is not ISO, otherwise, when a solution can be found no conclusions can be drawn as far as ISO is concerned.

Theorem 1 *Given a bounded net system $\mathscr{S}$, an unobservable set of transitions $T_{uo} \subset T$ and an integer $J \geq \mathscr{J}_{\min}$, let $\mathscr{M}_s$ be the set of secret initial markings.*

The system is not ISO if, for at least one secret marking $\vec{m}_s \in \mathscr{M}_s$, there exists an integer ρ such that the following feasibility problem

$$
\mathscr{X}\left(\vec{m}_{s_i},\rho\right):\left\{
\begin{array}{ll}
\vec{m}_s \geq \mathbf{Pre}_{uo}\cdot\vec{\epsilon}_{s_1} & \\
\vec{m}_s + \boldsymbol{C}_{uo}\cdot\vec{\epsilon}_{s_1} \geq \mathbf{Pre}_o\cdot\vec{s}_1 & \\
\vec{m}_s + \boldsymbol{C}_{uo}\cdot\vec{\epsilon}_{s_1} + \boldsymbol{C}_o\cdot\vec{s}_1 \geq \mathbf{Pre}_{uo}\cdot\vec{\epsilon}_{s_2} & \\
\ldots & (5a)\\
\vec{m}_s + \boldsymbol{C}_{uo}\cdot\sum_{i=1}^{J}\vec{\epsilon}_{s_i} + \boldsymbol{C}_o\cdot\sum_{i=1}^{J-1}\vec{s}_i \geq \mathbf{Pre}_o\cdot\vec{s}_J & \\
\boldsymbol{\nu} \geq \mathbf{Pre}_{uo}\cdot\vec{\epsilon}_{ns_1} & \\
\boldsymbol{\nu} + \boldsymbol{C}_{uo}\cdot\vec{\epsilon}_{ns_1} \geq \mathbf{Pre}_o\cdot\vec{s}_1 & \\
\boldsymbol{\nu} + \boldsymbol{C}_{uo}\cdot\vec{\epsilon}_{ns_1} + \boldsymbol{C}_o\cdot\vec{s}_1 \geq \mathbf{Pre}_{uo}\cdot\vec{\epsilon}_{ns_2} & \\
\ldots & (5b)\\
\boldsymbol{\nu} + \boldsymbol{C}_{uo}\cdot\sum_{i=1}^{J}\vec{\epsilon}_{ns_i} + \boldsymbol{C}_o\cdot\sum_{i=1}^{J-1}\vec{s}_i \geq \mathbf{Pre}_o\cdot\vec{s}_k & \\
\sum_{i=1}^{J}\sum_{j=1}^{n}\vec{s}_i(j) \geq \rho & (5c)\\
\boldsymbol{\nu} = \sum_{j=1}^{\mathrm{card}(\mathscr{M}_{ns})}\vec{m}_{ns_j}\circ(k_j\cdot\mathbf{1}), & (5d)\\
\sum_{j=1}^{\mathrm{card}(\mathscr{M}_{ns})}k_j = 1 & (5e)\\
\vec{\epsilon}_{s_i},\ \vec{\epsilon}_{ns_i},\ \vec{s}_i \in \mathbb{N}^n, \quad i=1,2,\ldots,J & (5f)\\
k_j \in \{0,1\}, \quad j=1,\ldots,\mathrm{card}(\mathscr{M}_{ns}) & (5g)
\end{array}\right.
$$

does not admit a solution, where "$\circ$" indicates the Hadamard product.

Proof First of all, let note that due to constraints (5d), (5e) and (5g), the vector $\boldsymbol{\nu}$ permits to *automatically* select a non-secret marking such that a solution to problem (5) exists. Hence, if (5) does not admit a solution, it means that for a given $\vec{m}_s \in \mathscr{M}_s$, the constraints are not feasible for any $\vec{m}_{ns} \in \mathscr{M}_{ns}$.

Moreover, in problem (5) three set of firing count vectors are included, namely

- $\vec{s}_i$ to model the firing of the observable transitions (see Remark 2);
- $\vec{\epsilon}_{s_i}$ to model the firing of unobservable transitions when the initial marking is the secret one, i.e. $\vec{m}_0 = \vec{m}_s$;
- $\vec{\epsilon}_{ns_i}$ to model the firing of unobservable transitions when the initial marking is a not-secret one, i.e. $\vec{m}_0 \in \mathscr{M}_{ns}$.

Therefore, taking into account also the constraint (5c), if the feasibility problem (5) does not admits a solution, then it means that there does not exist any firing count vector corresponding to a sufficiently long sequence that is enabled both from $\overrightarrow{m}_s$ and from any non secret marking in $\mathscr{M}_{ns}$, such that the observable transitions are mapped into the same firing count vectors $\overrightarrow{s}_i$. If this is the case, it means that condition (4) holds, implying the system to be not ISO.

Sufficiency of the condition given by Theorem 1 comes from the fact that in the firing count vectors we last the order of the firing of the transitions. Hence, if problem (5) admits a solution, this does not necessarily imply that condition (4) does not hold.

4 Examples

In this section we show the effectiveness of the proposed approach by applying the condition of Theorem 1 to some simple examples. All the considered ILP problems have been solved by using the GLPK solver [1].

Let us first consider the PN system in Fig. 1 with three transitions, two of which are unobservable, i.e. $T_{uo} = \{uo_1, uo_2\}$. Let us now consider the secrete marking set equal to $\mathscr{M}_s = \{\overrightarrow{m}_s\}$ with $\overrightarrow{m}_s = (2\ 0\ 0)^T$, and $\mathscr{M}_{ns} = \{\overrightarrow{m}_{ns}\}$ with $\overrightarrow{m}_{ns} = (0\ 2\ 0)^T$. for this simple example, it can be readily noticed that, with this choice for $\mathscr{M}_s$ and $\mathscr{M}_{ns}$ the net system is ISO; indeed there is only one observable transition, therefore an external intruder can only see the firing of transition o_1, which fires infinite times, regardless of the initial marking.

If we check the condition of Theorem 1 to assess the opacity of the considered example, by letting $\mathscr{J} = 2$ it follows that the feasibility problem generated finds a solution whatever the choice of ρ. Therefore, the condition of Theorem 1 does not hold. However, this is expected, being the system ISO and the proposed condition a sufficient one to assess if the system would have been not ISO.

Let us now consider the net in Fig. 2 which has five transitions, three of which are unobservable, i.e. $T_{uo} = \{uo_1, uo_2, uo_3\}$.

For this system we consider the following two cases:

(1) Two secret initial markings, $\overrightarrow{m}_{s_1} = (1\,2\,1\,0\,0)^T$ and $\overrightarrow{m}_{s_2} = (2\,2\,0\,0\,0)^T$, and the non secret one $\overrightarrow{m}_{ns_1} = (0\,2\,2\,0\,0)^T$, are considered.
Therefore, when applying the condition of Theorem 1 two feasibility problems should be solved. When $\overrightarrow{m}_{s_1}$ and $\overrightarrow{m}_{ns_1}$ are considered, the only observable transition which can fire is o_2. Therefore a solution to problem (5) can be found, since a possible firing sequence enabled from $\overrightarrow{m}_{s_1}$ is:

$$\sigma = uo_2 o_2 uo_2 uo_3 o_2 uo_2 o_2 uo_2 uo_3 o_2 \ldots,$$

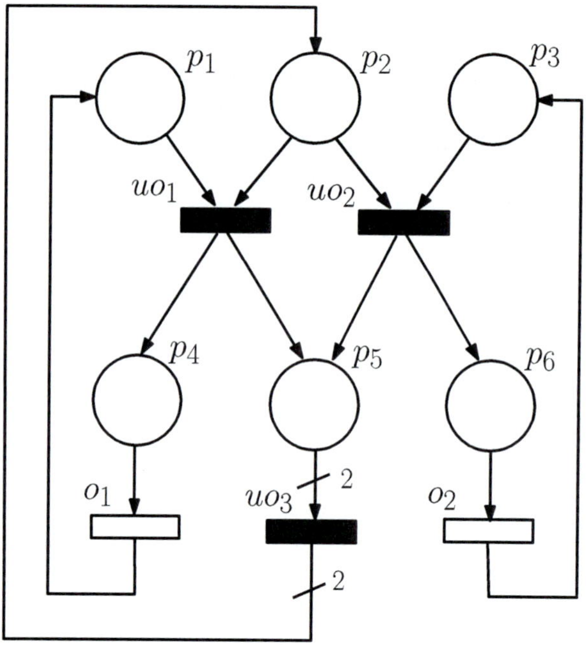

Fig. 2 Petri net system considered in Sect. 4

whose projection is:

$$Pr(\sigma) = o_2 o_2, \ldots .$$

On the other hand, if $\overrightarrow{m}_{s_2}$ is considered, then the feasibility problem's outcome highlights that there does not exist any firing sequence preventing o_1 from firing. As a matter of fact the feasibility problem fails to find a solution for $J = \rho = 4$, when $\overrightarrow{m}_{s_2}$ is considered, therefore the net is not ISO.

(2) In this second case, we considered one secret initial marking $\overrightarrow{m}_{s_1} = (2\ 2\ 0\ 0\ 0)^T$, and two non-secret initial markings $\overrightarrow{m}_{ns_1} = (0\ 2\ 2\ 0\ 0)^T$ and $\overrightarrow{m}_{ns_2} = (1\ 2\ 1\ 0\ 0)^T$.
Once again, we consider every sequence enabled from $\overrightarrow{m}_{s_1}$ and we conclude that the only observable transition which can fire is o_1.
While, similarly to the previous case, when m_{ns_1} is considered the firing of o_2 cannot be prevented for a sufficiently large ρ, the feasibility problem (2) admits a solution if $\overrightarrow{m}_{ns_2}$ is selected. Therefore, in this case, being the condition of Theorem 1 only sufficient, no conclusion can be drawn about the opacity of the system.

5 Conclusions

A sufficient condition to assess non initial state opacity in DES modeled as Petri net systems have been given. In particular, the proposed condition is based on the solution of ILP problems. Such an optimization-based approach can benefit by the use of off-the-shelf optimization tools, rather than relying on *ad hoc* algorithms, as in the case of graph-based approach (see also the comparison made in [3]).

Being only sufficient, the proposed result can be considered as a preliminary one, and possible line of future research consists in developing a set of constraints able to validate the necessary and sufficient conditions to assess ISO. Such an extension would enable to extend the supervisory control approach presented in [6] to enforce opacity in PN systems.

References

1. GLPK (GNU Linear Programming Kit): https://www.gnu.org/software/glpk/ (2021). Accessed 12.01.2021
2. Alladi, T., Chamola, V., Zeadally, S.: Industrial control systems: Cyberattack trends and countermeasures. Computer Communications **155**, 1–8 (2020)
3. Basile, F., Boussif, A., De Tommasi, G., Ghazel, M., Sterle, C.: Efficient diagnosability assessment via ILP optimization: a railway benchmark. In: 23rd IEEE International Conference on Emerging Technologies and Factory Automation, pp. 441–448, Torino, Italy, September 2018
4. Basile, F., Chiacchio, P., De Tommasi, G.: On $\mathcal{K}$-diagnosability of Petri nets via integer linear programming. Automatica **48**(9), 2047–2058 (2012)
5. Basile, F., De Tommasi, G.: An algebraic characterization of language-based opacity in labeled Petri nets. IFAC-PapersOnLine **51**(7), 329–336 (2018)
6. Basile, F., De Tommasi, G., Sterle, C.: Non-interference enforcement via supervisory control in bounded Petri nets. IEEE Trans. Autom. Control **66**(8), 3653–3666 (2020)
7. Busi, N., Gorrieri, R.: A survey on non-interference with Petri nets. In: Lectures on Concurrency and Petri Nets, pp. 328–344 (2004)
8. Cabasino, M.P., Giua, A., Seatzu, C.: Introduction to Petri nets. Control of Discrete-Event Systems, pp. 191–211. Springer (2013)
9. Cao, L., et al.: A survey of network attacks on cyber-physical systems. IEEE Access **8**, 44219–44227 (2020)
10. Cassandras, C.G., Lafortune, S.: Introduction to Discrete Event Systems, 2nd edn. Springer (2008)
11. García Vallés, F.: Contributions to the Structural and Symbolic Analysis of Place/Transition Nets with Applications to Flexible Manufacturing Systems and Asynchronous Circuits. Ph.D. thesis, Departamento de Informática e Ingeníeria de Sistemas, Centro Politecnico Superior, Universidad de Zaragoza, 1999
12. Jacob, R., Lesage, J.-J., Faure, J.-M.: Overview of discrete event systems opacity: Models, validation, and quantification. Annu. Rev. Control **41**, 135–146 (2016)
13. Keroglou, C., Lafortune, S.: Embedded insertion functions for opacity enforcement. IEEE Trans. Autom. Control **66**(9), 4184–4191 (2020)
14. Rashidinejad, A., et al.: Supervisory control of discrete-event systems under attacks: an overview and outlook. In: 2019 18th European Control Conference (ECC), pp. 1732–1739, Naples, Italy, June 2019

15. Saboori, A., Hadjicostis, C.N.: Opacity-enforcing supervisory strategies via state estimator constructions. IEEE Trans. Autom. Control **57**(5), 1155–1165 (2011)
16. Tong, Y., Li, Z., Seatzu, C., Giua, A.: Decidability of opacity verification problems in labeled petri net systems. Automatica **80**, 48–53 (2017)
17. Wu, Y.-C., Lafortune, S.: Comparative analysis of related notions of opacity in centralized and coordinated architectures. Discrete Event Dyn. Syst. **23**(3), 307–339 (2013)

Eco-Driving Adaptive Cruise Control via Model Predictive Control Enhanced with Improved Grey Wolf Optimization Algorithm

Raffaele Cappiello, Fabrizio Di Rosa, Alberto Petrillo, and Stefania Santini

Abstract In this paper, we suggest a novel Ecological Adaptive Cruise Control (Eco-ACC) system for an autonomous electric vehicle able to drive its motion while minimizing as much as possible its energy consumption. To this aim, we consider a Nonlinear Model Predictive Control (NMPC) method enhanced with an off-line Computational-intelligence (CI)-based optimization algorithm, i,e. the Improved-Grey Wolf Optimizer (I-GWO). Specifically, since the control performances strongly depend on the proper selection of the NMPC cost function, we propose the I-GWO algorithm to help the control designer find the sub-optimal weighting factors of the dynamic cost function optimized via the NMPC. An extensive numerical analysis involving realistic vehicle dynamics and a real-life Italian road network route confirm the effectiveness of the proposed approach in guaranteeing the ACC control objectives while ensuring energy saving.

Keywords Improved grey wolf optimization algorithm · MPC · ADAS · Adaptive cruise control

1 Introduction

In the last decades, autonomous driving has gained greater attention due to the benefit it could lead in terms of road safety since, as shown in [5, 13, 16, 17], the 90% of the traffic accidents are caused by human errors. To face this issue Advanced driving assistant systems (ADAS) plays a key role in improving driving safety and

Authors are in alphabetic order.

R. Cappiello F. Di Rosa · A. Petrillo (✉) · S. Santini
Department of Information Technology and Electrical Engineering, University of Napoli Federico II, Napoli, Italy
e-mail: raff.cappiello@studenti.unina.it; fa.dirosa@studenti.unina.it; alberto.petrillo@unina.it; stefania.santini@unina.it

A. Masone et al. (eds.), *Optimization and Data Science: Trends and Applications*, AIRO Springer Series 6, https://doi.org/10.1007/978-3-030-86286-2_11

comfort. Adaptive cruise control (ACC), whose aim is to maintain an Ego vehicle at a certain distance w.r.t. the preceding one, represents one of the main technology that helps in this direction [23]. Besides road safety, the environmental issue has received considerable attention since road transport is responsible for 16.5% of the global greenhouse gas emissions [3]. With the aim at reducing emissions, spreading attention has been given to hybrid/electric vehicles, where, however, the battery usage management strongly affect the vehicle life cycle [22].

In the technical literature, due to its capabilities to reach multiple control objectives while handling multiple constraints in a receding horizon fashion, Model Predictive control approaches are widely proposed in [8, 11, 24] to design Ecological-ACC. The choice of both cost function and the weightings factors strongly influences the closed-loop system's performance, and the energy-saving requirements [1]. Indeed, the tuning of the weightings parameters is challenging as they are related to the closed-loop performance in a complex manner [21]. However, this problem is not considered in the above-mentioned work, and the tuning parameters are selected via a try and error process. To overcome this problem, in this work, we propose a novel Eco-ACC control system which combines the Nonlinear Model Predictive Control (NMPC) and computational intelligence (CI)-based optimization algorithm, i.e. the improved Grey Wolf algorithm (I-GWO) [4], for the sub-optimal tuning of the weightings parameters w.r.t. the closed-loop performances. Specifically, since the vehicle dynamic is inherent nonlinear (air drag, gravity force, rolling resistance, etc...) and strongly affected by uncertainty (e.g. due to road shape and slope), an on-line Nonlinear Model Predictive Control (NMPC) approach is used to keep the accuracy of the prediction model faithful to reality. Conversely, I-GWO works off-line based on a data set derived by running several simulations of the behaviour of the autonomous electric vehicles under the action of the ACC control system. Then, by evaluating a properly designed cost function, the I-GWO optimization algorithm finds the best sub-optimal solution for the NMPC tuning parameters. The obtained result is then used to shape the NMPC cost function and solve the on-line NMPC control problem. To validate the proposed control approach, an extensive numerical validation has been carried out considering a real Italian road section (i.e. the route *Roma-Padova*). Moreover, to perform an adequate analysis, a comparison w.r.t an NMPC control strategy whose weightings parameters have been chosen via try and error technique has been performed. To sum up, numerical results confirm the effectiveness of the approach and its efficiency in ensuring energy saving.

Finally, the paper is organized as follows. In Sect. 2, the grey wolf algorithm, as well as the improved one is introduced. Section 3 presents the problem statement and control-oriented vehicle models. In Sect. 4, we suggest the proposed control approach while in Sect. 5, the numerical analysis is carried out. In Sect. 6, conclusions are drawn.

2 Mathematical Preliminaries

2.1 *Grey Wolf Optimizer*

The main idea that inspired the author of the grey wolf optimizer is the behaviour, during the hunting, of grey wolves in nature. In general, the hierarchy of grey wolves is structured as follow [20]: (i) Alfa wolves (α) are responsible for nearly all group decisions such as time to hunting, places to stay and, time to wake up; (ii) Beta wolves (β) is the second wolf in the hierarchy and helps the alpha to make decisions; (iii) Delta wolves (δ) is the lowest rank of grey wolves' social hierarchy, but delta wolves are dominant compared with omega wolves; (iv) Omega wolves (ω) are the remaining wolves that can not be classified as alfa, beta, or delta.

2.1.1 Grey Wolf Optimizer: Principle of Operation

The hunting of the grey wolves consists of three main step: (i) encircling; (ii) hunting; (iii) attacking the prey.

2.1.2 Encircling

In order to simulate and model the encircling behaviour of wolves the following equation are used:

$$\vec{D} = \left|\vec{C} \cdot \vec{X}_p(t) - \vec{X}(t)\right|, \quad \vec{X}(t+1) = \vec{X}_p(t) - \vec{A} \cdot \vec{D} \tag{1}$$

where t defines the current iteration $\vec{A}$ and $\vec{C}$ are coefficient vectors calculated as in (2), X_p is the position of the prey, and $\vec{X}$ indicate the position vector of a grey wolf.

$$\vec{A} = 2\vec{a} \cdot \vec{r}_1 - \vec{a}, \quad \vec{C} = 2 \cdot \vec{r}_2 \tag{2}$$

where $\vec{a}$ components are linearly decreased from 2 to 0 over the iterations as in (3) and r_1 and r_2 are random vectors in $[0, 1]$.

$$a(t) = 2 - (2 \times t)/\text{MaxIter} \tag{3}$$

2.1.3 Hunting

As the wolves usually operate in a natural environment, they know where the prey is and where they should go to encircle the prey. Instead, in our scenario, the wolves do not know where is the prey, defined as $\vec{X}_p$ in (1), and the central assumption of the algorithm is that the main three wolves, *alpha*, *beta*, and *delta*, have a better knowledge about the current and exact position of the prey. Other all the wolves follow the previously obtained best solution during the optimization process. In order to model what is said, the following equations are used.

$$\vec{D}_{\alpha} = \left|\vec{C}_1 \cdot \vec{X}_{\alpha} - \vec{X}\right|; \quad \vec{D}_{\beta} = \left|\vec{C}_2 \cdot \vec{X}_{\beta} - \vec{X}\right|; \quad \vec{D}_{\delta} = \left|\vec{C}_3 \cdot \vec{X}_{\delta} - \vec{X}\right| \tag{4}$$

$$\vec{X}_{i,1} = \vec{X}_{\alpha} - \vec{A}_{i,1} \cdot \left(\vec{D}_{\alpha}\right); \quad \vec{X}_{i,2} = \vec{X}_{\beta} - \vec{A}_{i,2} \cdot \left(\vec{D}_{\beta}\right); \quad \vec{X}_{i,3} = \vec{X}_{\delta} - \vec{A}_{i,3} \cdot \left(\vec{D}_{\delta}\right) \tag{5}$$

$$\vec{X}_{i-GWO}(t+1) = \frac{\vec{X}_{i,1} + \vec{X}_{i,2} + \vec{X}_{i,3}}{3} \tag{6}$$

where $\vec{X}_{\alpha}$, $\vec{X}_{\beta}$, $\vec{X}_{\delta}$ are the first three best solutions at iteration t, $\vec{A}_{i,1}$, $\vec{A}_{i,2}$, $\vec{A}_{i,3}$, are calculated as in (2).

It can be observed that the final position of the i-th wolf i.e. $\vec{X}_{i-GWO}(t+1)$ would be, due to the presence of $\vec{A}_i$, in a random place within a circle, defined by the positions of *alfa*, *beta*, and *delta* in the search space.

2.1.4 Attacking

The hunting process is terminated when the prey stops moving, and wolves start an attack. This can be done mathematically by the value of a which is linearly decreased over the curse of iterations controlling the exploration and exploitation as in (3).

2.2 *Improved Grey Wolf Optimizer: IGWO*

In a classical GWO approach, the searching is driven by the α, β and δ wolves, which promise to find the best solution. Unfortunately, this behaviour may lead to entrapment in locally optimal solution and reduce the population's diversity. An improved version, the Improved Grey Wolf Optimizer (I-GWO), has been lately developed by Mirjalili et al. [14] to overcome these common problems. The I-GWO adds a movement strategy called dimension learning based hunting (DLH). The DLH consists of taking into account neighbouring wolves when updating each wolf's position in the pack. For all iteration t a neighbourhood $N_i(t)$ for each wolf $X_i(t)$ is defined as in (7).

$$N_i(t) = \left\{X_j(t) \mid D_i\left(X_i(t), X_j(t)\right) \leqslant R_i(t), X_j(t) \in \text{Pack}\right\} \tag{7}$$

where $R_i(t)$ is calculated as in (8) using Euclidean distance between the current position of $X_i(t)$ and the candidate position $X_{i-GWO}(t+1)$ defined in (6) and $D_i(\cdot)$ is Euclidean distance between $X_i(t)$ and $X_j(t)$ for all i, j in the pack.

$$R_i(t) = \|X_i(t) - X_{i-Gwo}(t+1)\| \tag{8}$$

Once a neighbourhood of $X_i(t)$ is constructed, the position of the i-th wolf at the next iteration, according to the DHL approach, is calculated:

$$X_{i-DLH}(t+1) = X_i(t) + \text{rand} \times (X_n(t) - X_r(t)) \tag{9}$$

where $X_i(t)$ is the position of the wolf in analysis at the current iteration, $X_n(t)$ is a random wolf selected between the neighbors $N_i(t)$ and $X_r(t)$ is a random wolf from the whole pack.

At the end of this phase, there are two possible positions for the $X_i(t+1)$ wolf, one calculated via the classical GWO approach X_{i-GWO} and the other one obtained by the DHL technique X_{i-DLH}. To chose the effective position for the for the i-th wolf componing the pack at the next iteration the criteria described in (10) is used.

$$X_i(t+1) = \begin{cases} X_{i-GWO}(t+1), & \text{if} f(X_{i-GWO}) < f(X_{i-DLH}) \\ X_{i-DLH}(t+1) & \text{otherwise} \end{cases} \tag{10}$$

where $f(X_j)$ is the fitness value for the generic X_j position, X_{i-GWO} and X_{i-DLH} are respectively evaluated as in (6)–(9).

3 Problem Statement

Consider an autonomous electric vehicle, called Ego vehicle, equipped with on-board proximity sensors (such as radar, lidar and camera) and with on-board control unit that allows the vehicle to move into its surrounding traffic scenario autonomously. Our aim is to design a novel Ecological Adaptive cruise Control (Eco-ACC) that allows maintaining an optimal inter-vehicle distance from the preceding vehicle while minimizing energy consumption to reach the control objective (see Fig. 1). Specifically, the proposed control system has to pursue the following multiple control objectives: (i) tracking of the preceding vehicle speed profile; (ii) maintaining a desired safe inter-vehicle distance to avoid any possible

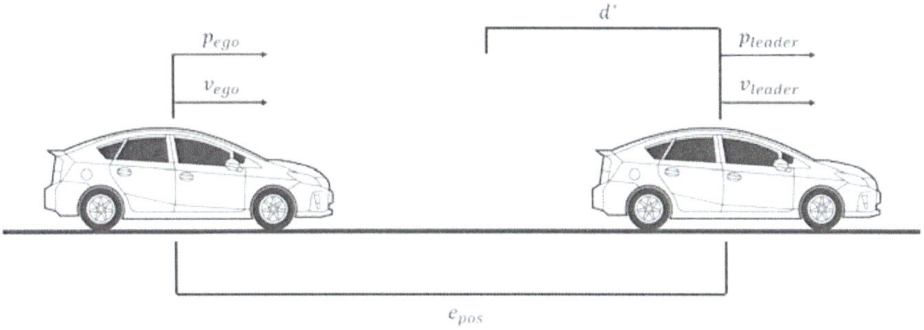

Fig. 1 Adaptive cruise control system

collisions; (iii) ensuring a safe and comfortable driving experience; (iv) minimizing energy consumption and, hence, ensuring improved battery management.

3.1 Electric Autonomous Ego Vehicle

The Ego Vehicle behaviour is described by its longitudinal motion, which takes into account the aerodynamics drag, rolling resistance, and gravitational force [18]. To derive a mathematical model for control design, we assume that the driving/braking torques are integrated into one high-level control input. Accordingly, the motion of the ego vehicle can be described by the following non-linear dynamical system [19, 25]:

$$\begin{bmatrix} \dot{p}_{ego}(t) \\ \dot{v}_{ego}(t) \end{bmatrix} = \begin{bmatrix} v_{ego}(t) \\ -\frac{1}{2m} C_d C_h \rho_a A v_{ego}^2(t) - \mu g \cos(\theta(t)) - g \sin(\theta(t)) + \frac{\eta}{R \cdot m} u(t) \end{bmatrix} \quad (11)$$

where $p_{ego}(t)$[m] $\in \mathbb{R}$ and $v_{ego}(t)$ [m/s] $\in \mathbb{R}$ are the position and the velocity of the Ego vehicle w.r.t. the reference road frame, respectively; $u(t)$ [N m] is the control input that represents the vehicle propulsion torque, i.e. the driving/braking torque; m [kg] is the vehicle mass; η is the drive-train mechanical efficiency; R [m] is the wheel radius; C_d is the vehicle drag coefficient; ρ_a [kg/m^3] is the air density; A [m^2] is the vehicle frontal area; C_h is a correction factor (unitless) due to the altitude of the road where vehicle moves; μ is the rolling resistance coefficient; g [m/s^2] is the gravity acceleration while $\theta(t)$ [rad] is the road-track slope. According to [19], the altitude correction factor is computed as $1 - 0.085H$ where H [km] is the road altitude. For sake of compact notation, by introducing the state vector $x(t) = [p_{ego}(t)\ v_{ego}(t)]^T$, we can recast the vehicle dynamics as in (11) as

$$\dot{x}(t) = f(x(t), u(t)) \quad (12)$$

For the EV battery pack model [7, 12] we consider an equivalent simplified electric circuit consisting of an internal voltage source V_{oc} and one resistance R_b. The voltage at the terminal of the battery is computed as:

$$V_b(t) = V_{oc} - R_b I_b(t) \quad (13)$$

being I_b the battery current. Indicating with $P_{req}(t)$ the power required at the battery, $I_b(t)$ can be derived as

$$I_b(t) = \frac{V_{oc} - \sqrt{V_{oc}^2 - \frac{4R_b P_{req}(t)}{n_b}}}{2R_b} \quad (14)$$

being n_b the number of cells constituting the battery. The battery State of Charge can be hence computed as:

$$SOC(t) = -\frac{1}{C_{batt}} \int_0^t I_b(\tau) d\tau \quad (15)$$

where C_{batt} is the battery capacity. Finally, the power required to the battery P_{req} is evaluated as [15]

$$P_{req}(t) = \omega(t)u(t) + \frac{R_m}{K^2}u(t)^2 + P_{aux} \tag{16}$$

where ω is the rotational speed of the DC motor derived from the linear speed as $\omega = \frac{v_{ego}(t)R_t}{R}$, being Rt the transmission ratio; $K = K_a\phi$, being K the armature constant and Φ (in Weber $[Wb]$) the armature magnetic flux; R_m is the DC motor resistance; P_{aux}, not determined by vehicle velocity and acceleration, is the auxiliary power loss including some electronic devices in the vehicle, such as radios, lights, air conditioners and so on [10].

3.2 Control Objectives

Given a reference behaviour as imposed by the preceding vehicle and measured via the proximity sensors, i.e. $p_{ref}(t)$ and $v_{ref}(t)$-being these latter the preceding vehicle position and speed, respectively- our aim is to design an Eco-ACC such that:

$$\lim_{t\to\infty} \left(p_{\text{ref}}(t) - p_{ego}(t) - d^\star\right) = 0 \tag{17}$$

$$\lim_{t\to\infty} \left(v_{\text{ref}}(t) - v_{ego}(t)\right) = 0 \tag{18}$$

$$p_{\text{ref}}(t) - p_{ego}(t) \geqslant d^\star \tag{19}$$

$$\min \int_0^\infty P_{req}(\tau)d\tau \tag{20}$$

where $d^\star$ the desired constant inter-vehicle distance between the ego vehicle and the preceding vehicle. Note that the control objectives (17)–(19) will refer to the tracking of a desired behaviour w.r.t. the preceding vehicle while the control goal (20) is related to the energy saving requirement.

4 Control Design

To fulfil the control objectives proposed in Sect. 3.2, in this section Eco-ACC system is designed according to an NMPC control methodology that leverages the meta-heuristic I-GWO optimization algorithm for the proper selection of the controller weight parameters.

4.1 Nonlinear Model Predictive Control Design

The NMPC is chosen due to the nonlinear electric vehicle dynamics, the variability of driving conditions and the multitude of physical constraints for involved dynamic variables, hence coping with the dynamical system uncertainties by anticipating future situations [2]. See Fig. 2 for the proposed control architecture.

Specifically, we design the control action $u(t)$ in (11) by solving the following multiple optimization problem:

$$\min_{u(t)} \int_{t}^{t+T} \mathcal{J}(x(\tau|t), u(\tau|t))d\tau \tag{21}$$

subject to:

$$\dot{x}(t) = f(x(t), u(t)) \tag{22a}$$

$$a_{\min} \leq \dot{v}_{ego}(\tau|t) \leq a_{\max} \tag{22b}$$

$$j_{\min} \leq \ddot{v}_{ego}(\tau|t) \leq j_{\max} \tag{22c}$$

$$\frac{a_{min} Rm}{\eta} \leq u(\tau|t) \leq \frac{a_{max} Rm}{\eta} \tag{22d}$$

where T is the prediction horizon; $x_i(\tau|t)$ and $u_i(\tau|t)$ are the prediction of the state variable and the control input respectively of the dynamical model (12); $a_{\min}$ and $a_{\max}$ denote the minimum and maximum bounds for the Ego vehicle acceleration variable; $j_{\min}$ and $j_{\max}$ denote the minimum and maximum bounds for the Ego

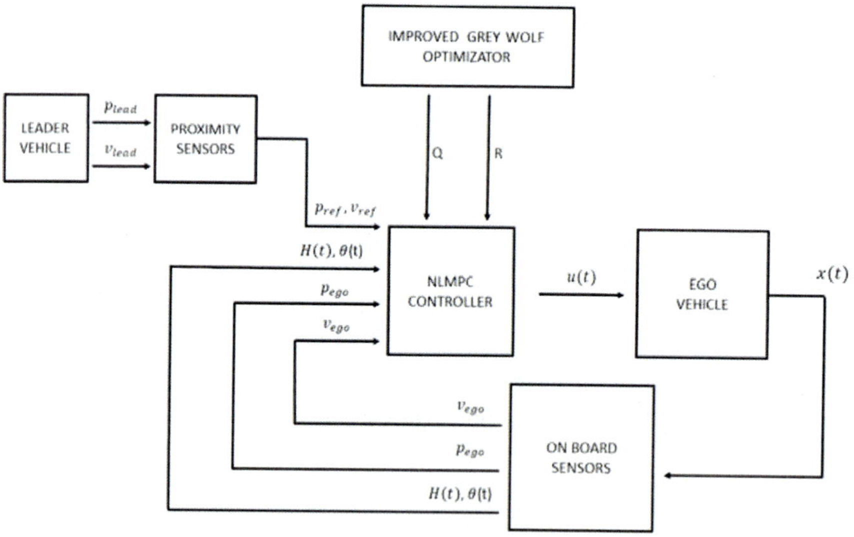

Fig. 2 Control architecture for the Ego vehicle

vehicle jerk variable. Moreover, the cost function $\mathscr{J}$ is defined as:

$$\mathscr{J} = \omega_1 \| p_{\text{ref}} - p_{ego} - d^\star \|^2 + \omega_2 \| v_{\text{ref}} - v_{ego} \|^2 + \omega_3 \| u \|^2 \tag{23}$$

where ω_1, ω_2 and ω_3 are non-negative weighting scalar which are tuned by exploiting the I-GWO optimization algorithm.

Note that, the cost function as in (23) is selected in order to satisfy the multi-control objectives as in Sect. 3.2, while the constraints (22b)–(22d) are imposed so to guarantee a safe and comfortable driving experience.

4.2 *Grey Wolf Optimization Algorithm for the Tuning of the NMPC Weights*

Typically, the weighting factors ω_ρ ($\rho = 1, 2, 3$) into the cost function to be optimized as in (23) are chosen using trial and error techniques [1]. However, this choice strongly influences the closed-loop performances and the effectiveness of the proposed control strategy. To face this issue, herein we propose the exploitation of the offline meta-heuristic optimization algorithm, i.e. the I-GWO, to obtain a sub-optimal trade-off among the different weighting factors ω_k and hence among the different control goals as detailed in Sect. 3.2. To find the sub-optimal values of ω_ρ ($\rho = 1, 2, 3$), called $\omega_\rho^\star$ ($\rho = 1, 2, 3$) are obtained according to Algorithm 1.

More specifically, after initialising each search agent in the pack with three random positions within the bounded search region, we run the NMPC with the selected value, and we evaluate the following cost function

$$J = \sum_{\forall k} \psi_1 k \frac{(e_{pos}(k) - d^\star)^2}{\eta_1} + \psi_2 k \frac{(e_{vel}(k))^2}{\eta_2} + \psi_3 \frac{SOC(k)}{\eta_3} \tag{24}$$

$e_{pos} = p_{ref}(k) - p_{ego}(k)$ is the position error, $e_{vel} = v_{ref}(k) - v_{ego}(k)$ is the speed error, k is the generic simulation step and η_ρ are weighting factors which allows normalizing each term of J, and hence equally weighting each of them.

After running a number of simulations equal to the number of wolves that compose the pack, the best three solutions are combined to estimate the position of the optimum, which corresponds in our case to the three weights that minimise the cost (24). At the following iteration, the wolves' positions are updated using the assumed position of the optimal as a "central point" and information about a defined neighbourhood (10). The process is then repeated for a number of times equal to the maximum number of iterations, at the end of which three optimal sub-solutions are obtained and denoted as ω_1^*, ω_2^* and ω_3^*.

Algorithm 1 Computing the Optimal NMPC weighting factors ω_ρ $(\rho = 1, 2, 3)$

Declarations:
Let $e_{pos} = (p_{ref} - p_{ego} - d^\star)$ the position error;
Let $e_{vel} = (v_{ref} - v_{ego})$ the velocity error;
Let SOC the value of state of charge;
Let $\alpha = [\alpha_1, \alpha_2, \alpha_3]$ the alpha grey wolf, the first best solution;
Let $\beta = [\beta_1, \beta_2, \beta_3]$ the beta grey wolf, the second best solution;
Let $\gamma = [\gamma_1, \gamma_2, \gamma_3]$ the alpha grey wolf, the third best solution;
Initialization:
Initialize the number of grey wolf as $n = 55$;
Initialize the number of max iteration as 15;
Initialize the population of grey wolves $X_i = (i = 1, 2, \ldots, n)$
Set $t = 0$
Algorithm:
while $t <$ *Max number of iteration* **do**
  for *each search agent* **do**
    if *t==0* **then**
      Randomically initialize the position of the *search agent* in search region;
    else
      Update position of the *search agent* based on the three best previous solutions and neighbourhood as in (10);
    end
    Start the simulation;
    Get data from simulation and evaluate and evaluate e_{pos}, e_{vel} and SOC;
    Evaluate the cost function for the I-GWO algorithm as in (24);
  end
  Find the best three solutions: α, β and γ according to (10);
  t=t+1;
end
$\omega_1^* = \alpha(1)$;
$\omega_2^* = \alpha(2)$;
$\omega_3^* = \alpha(3)$;

5 Numerical Analysis

In this section, we disclose the effectiveness of the proposed NMPC strategy enhanced with the I-GWO optimization algorithm to ensure that the ego vehicle tracks the reference behaviour imposed by the preceding vehicle while optimizing its energy consumption. Numerical simulations are carried out by exploiting the Matlab/Simulink simulation platform considering the *Roma-Padova* Italian road network whose altitude and road profile are reported in Fig.3. The values of the EGO vehicle motion parameters [9], as well as the ones related to the battery and the electric motor [6] are in Table 1. Regarding the reference behaviour to be imposed on the Ego vehicle, we consider that the preceding vehicle moves with the constant speed of $v_{ref} = 15$ [m/s] while the desired inter-vehicle distance between the Ego

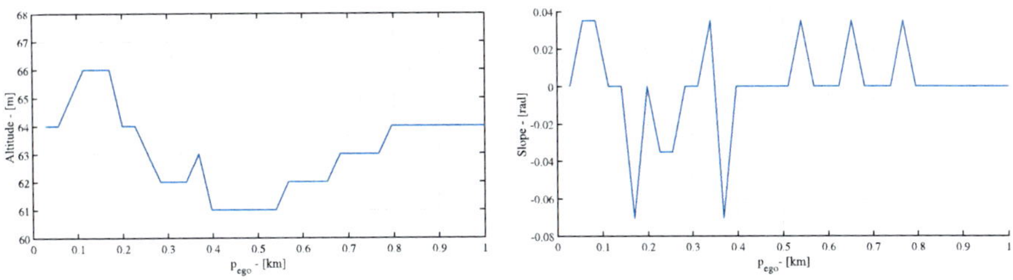

Fig. 3 Altitude and slope profile for the appraised *Roma-Padova* road network

vehicle and the preceding is selected as $d^\star = 10$ [m]. Finally, to show the benefits of the proposed control approach we compare the performances achieved with our I-GWO-enhanced NMPC with the ones achievable with a classical NMPC where the weighting factors are not optimized.

5.1 Numerical Results

By leveraging the I-GWO optimization procedure as in Algorithm 1, we derive the NMPC control sub-optimal weights as $\omega_1^* = 20.6537$, $\omega_2^* = 1.74043$ and $\omega_3^* = 0.0749339$, found in 12 iterations as shown in Fig. 4b. Moreover, in Fig. 4a, we plot the evolution of (24) as a function of the pair ω_1, ω_2 for the optimal value of $\omega_3^\star$ where the sub-optimal point is also explicitly highlighted. Numerical results in Fig. 5 confirm the proposed control approach's effectiveness. Indeed, the Ego vehicle, starting from random initial conditions, reaches the constant speed as imposed by the leading vehicle in 20 s with an overshoot of 13% (see Fig. 5a) and maintains the inter-vehicle spacing of $d^\star$ w.r.t. the preceding vehicle itself. This confirms the control approach's effectiveness in guaranteeing the correct functioning of the proposed IGW-based ECO-NMPC ACC control system.

5.2 Comparison Analysis

Here we compare our approach's performances with a classical NMPC strategy where the weighting factors are not optimized with the proposed I-GWO but are selected via a try and error process, i.e. $\omega_1 = 23$; $\omega_2 = 1.97$; $\omega_3 = 0.06$. By comparing the performances in Fig. 6, it is possible to observe our approach's improved performances in ensuring an Eco-ACC control w.r.t. the controller in comparison. The speed the of vehicle controlled by NMPC, with weights selected via try and error, faster converges to the desired value but with a worst overshoot. This behaviour causes a less smooth dynamic behaviour which implies an increasing energy consumption. Indeed, by evaluating the integral of battery current necessary

Table 1 Ego vehicle parameters

m [kg]	η [−]	R [m]	μ [−]	C_D [−]	A [m^2]	a_{max} [m/s^2]	a_{min} [m/s^2]	n_b [−]	C_{batt} [Ah]	R_b [Ω]	R_t [−]	R_m [Ω]	P_{aux} [W]	K [Vs]	j_{min} [ms^{-3}]	j_{max} [ms^{-3}]
1521	0.93	0.5	0.020	0.28	2.3315	2.5	−6.0	192	40	0.01	0.95	0.11	750	10.08	−2	2

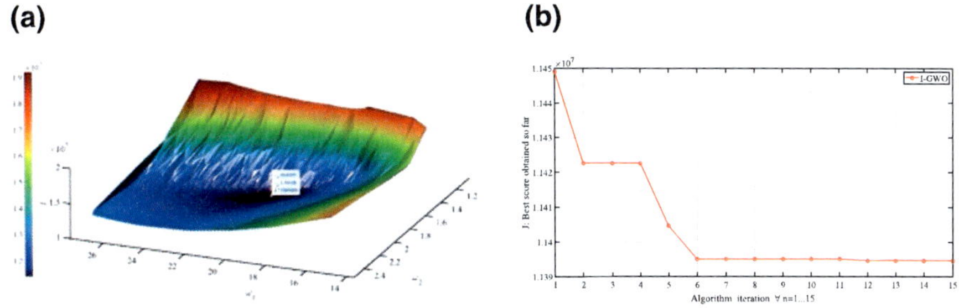

Fig. 4 I-GWO optimization algorithm results: (**a**) evaluation of the I-GWO cost function; (**b**) evolution of the optimal cost function vs iterations

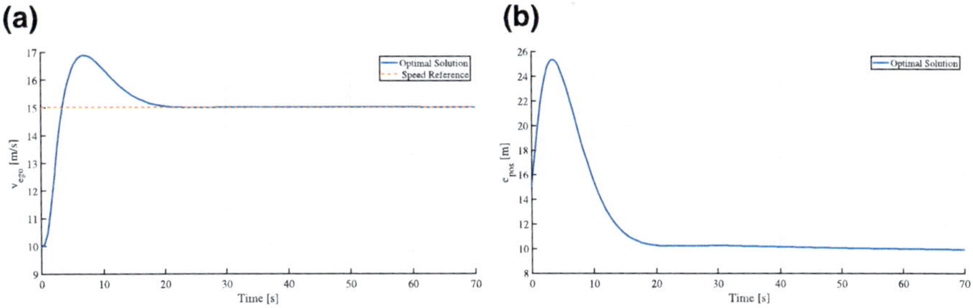

Fig. 5 Control performance with weighting factor optimized. Time history of: (**a**) Ego vehicle speed; (**b**) position error

to drive the electric vehicle, it is possible to observe how our approach can ensure an energy reduction of 2.5% (see Fig. 6c).

6 Conclusion

In this paper, we have proposed a novel Eco-ACC control system for autonomous electric vehicles that leverages the NMPC technique enhanced with the I-GWO meta-heuristic optimization algorithm for the proper selection of the weighting factors of the NMPC cost function. The proposed approach has aimed to drive the Ego vehicle according to an ACC control objective while minimizing as much as possible the energy consumption. Since the above-mentioned control objectives strongly depend on the proper choice of the cost function to be optimized, the I-GWO algorithm has helped design the NMPC problem via an off-line optimization of its weighting factors. Numerical analysis has confirmed the effectiveness of the proposed approach in guaranteeing the ACC control objectives while leading to considerable energy saving.

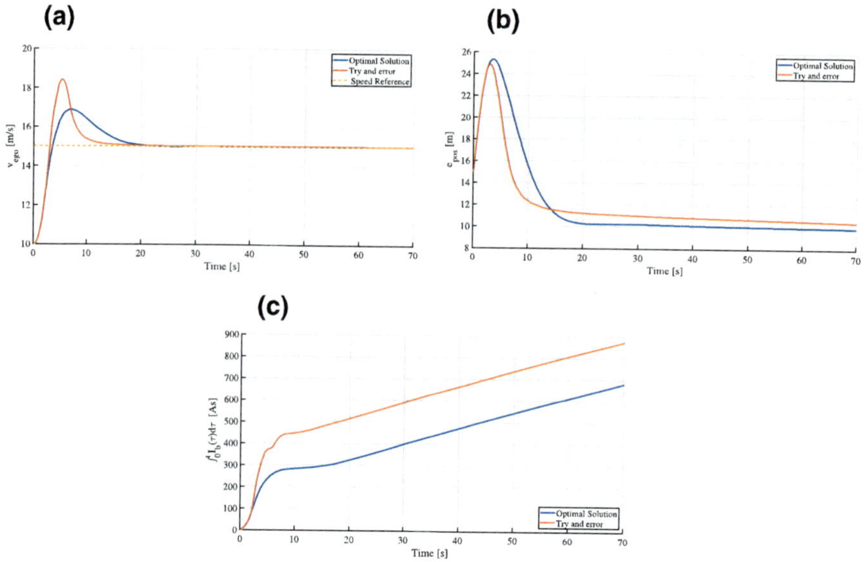

Fig. 6 Comparison between the proposed I-GWO-based NMPC controller and non-optimized NMPC controller. Time history of: (**a**) Ego vehicle speed; (**b**) Position error; (**c**) energy consumption $\int_0^t I_b(\tau)d\tau$

References

1. F. Allgöwer, A. Zheng. Nonlinear Model Predictive Control, vol. 26. Birkhäuser (2012)
2. Amodeo, M., Di Vaio, M., Petrillo, A., Salvi, A., Santini, S.: Optimization of fuel consumption and battery life cycle in a fleet of connected hybrid electric vehicles via distributed nonlinear model predictive control. In: 2018 European Control Conference (ECC), pp. 947–952. IEEE (2018)
3. Birol, F.: Co2 emissions from fuel combustion. International Energy Agency (2016)
4. Bozorg-Haddad, O.: Advanced Optimization by Nature-Inspired Algorithms. Springer (2018)
5. Fiengo, G., Lui, D.G., Petrillo, A., Santini, S., Tufo, M.: Distributed robust pid control for leader tracking in uncertain connected ground vehicles with v2v communication delay. IEEE/ASME Trans. Mechatron. **24**(3), 1153–1165 (2019)
6. He, X., Wu, X.: Eco-driving advisory strategies for a platoon of mixed gasoline and electric vehicles in a connected vehicle system. Transport. Res. D Transport Environ. **63**, 907–922 (2018)
7. Iannuzzi, D., Santini, S., Petrillo, A., Borrino, P.I.: Design optimization of electric kart for racing sport application. In: 2018 IEEE International Conference on Electrical Systems for Aircraft, Railway, Ship Propulsion and Road Vehicles International Transportation Electrification Conference (ESARS-ITEC), pp. 1–6 (2018)
8. Jia, Y., Jibrin, R., Gorges, D.: Energy-optimal adaptive cruise control for electric vehicles based on linear and nonlinear model predictive control. IEEE Trans. Veh. Technol. (2020)
9. Li, K., Gao, F., Li, S.E., Zheng, Y., Gao, H.: Robust cooperation of connected vehicle systems with eigenvalue-bounded interaction topologies in the presence of uncertain dynamics. Front. Mech. Eng. **13**(3), 354–367 (2018)

10. Li, Y., Zhang, L., Zheng, H., He, X., Peeta, S., Zheng, T., Li, Y.: Evaluating the energy consumption of electric vehicles based on car-following model under non-lane discipline. Nonlinear Dynamics **82**(1-2), 629–641 (2015)
11. Magdici, S., Althoff, M.: Adaptive cruise control with safety guarantees for autonomous vehicles. IFAC-PapersOnLine **50**(1), 5774–5781 (2017)
12. Maia, R., Silva, M., Araújo, R., Nunes, U.: Electrical vehicle modeling: A fuzzy logic model for regenerative braking. Expert Syst. Appl. **42**(22), 8504–8519 (2015)
13. Manfredi, S., Petrillo, A., Santini, S.: Distributed pi control for heterogeneous nonlinear platoon of autonomous connected vehicles. IFAC-PapersOnLine **53**(2), 15229–15234 (2020)
14. Nadimi-Shahraki, M.H., Taghian, S., Mirjalili, S.: An improved grey wolf optimizer for solving engineering problems. Expert Syst. Appl. **166**, 113917 (2021)
15. Petit, N., Sciarretta, A.: Optimal drive of electric vehicles using an inversion-based trajectory generation approach. IFAC Proc. Vol. **44**(1), 14519–14526 (2011)
16. Petrillo, A., Pescapé, A., Santini, S.: A secure adaptive control for cooperative driving of autonomous connected vehicles in the presence of heterogeneous communication delays and cyberattacks. IEEE Trans. Cybern. **51**(3), 1134–1149 (2021)
17. Petrillo, A., Salvi, A., Santini, S., Valente, A.S.: Adaptive multi-agents synchronization for collaborative driving of autonomous vehicles with multiple communication delays. Transport. Res. C Emerg. Technol. **86**, 372–392 (2018)
18. Rajamani, R.: Vehicle Dynamics and Control. Springer Science & Business Media (2011)
19. Rakha, H.A., Ahn, K., Moran, K., Saerens, B., Van den Bulck, E.: Virginia tech comprehensive power-based fuel consumption model: model development and testing. Transport. Res. D Transport Environ. **16**(7), 492–503 (2011)
20. Rezaei, H., Bozorg-Haddad, O., Chu, X.: Grey Wolf Optimization (GWO) Algorithm, pp. 81–91 (07 2018)
21. Shah, G., Engell, S.: Tuning mpc for desired closed-loop performance for mimo systems. In: Proceedings of the 2011 American Control Conference, pp. 4404–4409. IEEE (2011)
22. Tie, S.F., Tan, C.W.: A review of energy sources and energy management system in electric vehicles. Renew. Sustain. Energy Rev. **20**, 82–102 (2013)
23. Wang, Z., Wu, G., Barth, M.J.: A review on cooperative adaptive cruise control (cacc) systems: Architectures, controls, and applications. In: 2018 21st International Conference on Intelligent Transportation Systems (ITSC), pp. 2884–2891. IEEE (2018)
24. Weißmann, A., Görges, D., Lin, X.: Energy-optimal adaptive cruise control combining model predictive control and dynamic programming. Control Eng. Pract. **72**, 125–137 (2018)
25. Wu, Y., Li, S.E., Cortés, J., Poolla, K.: Distributed sliding mode control for nonlinear heterogeneous platoon systems with positive definite topologies. IEEE Trans. Control Syst. Technol. **28**(4), 1272–1283 (2019)

Part V
OR in Industry

Optimizing and Evaluating a Maintenance Strategy for Multi-Component Systems

Lucía Bautista Bárcena and Inmaculada T. Castro

Abstract Maintenance optimization is a key challenge in engineering and industry. Nowadays, the development of multi-component and complex systems has increased the difficulty of studying and analyzing these systems. Preventive and corrective thresholds are set to check the system deterioration. If one of those thresholds is exceeded by a component, this component is preventively or corrective maintained, depending on its degradation level. The system is periodically inspected, and these inspection times are used as opportunities to perform a preventive maintenance in the rest of the components. The optimal cost, minimizing the time between inspections and the preventive thresholds, is evaluated. Different techniques for the optimization process are employed; typical Monte-Carlo simulation combined with some meta-heuristic algorithms.

Keywords Maintenance · Reliability · Stochastic processes

1 Introduction

The evaluation of the maintenance of multi-component systems has been a key challenge in engineering and industry, since we can greatly reduce maintenance costs and avoid system breakdowns by applying the correct policy [1, 13, 15].

In this work, we are going to study a complex system consisting of two different types of components: monitored (or degrading) components and non-monitored (or non-degrading) components. Both types of components deteriorate following different processes: monitored components are subject to a continuous degradation process following a Gamma process, while non-monitored components fail at certain times, following a Poisson arrival process. These non-monitored components

L. B. Bárcena (✉) · I. T. Castro
Department of Mathematics, University of Extremadura, Badajoz, Spain
e-mail: luciabb@unex.es; inmatorres@unex.es

A. Masone et al. (eds.), *Optimization and Data Science: Trends and Applications*, AIRO Springer Series 6, https://doi.org/10.1007/978-3-030-86286-2_12

can only be maintained upon failure. This model was proposed in [10] and [4] with two blocks of units.

In real life we can find lots of systems with that structure: some pieces that degrade continuously (for example, a conveyor belt, whose degradation can be seen with the naked eye) and other ones that are subject to sudden failures and we cannot predict their failure, for example, a light bulb [4, 10].

This work is organized into sections. The problem is presented in the Introduction Sect. 1. The problem is modelled mathematically trough different probability distributions and the maintenance policy is explained in Sect. 2. In Sect. 3, the objective cost function is described, with a well-known theorem useful in this paper. Section 4 is devoted to the optimization algorithms used to optimize the system maintenance. Finally, a numerical example is presented in Sect. 5 with identical and non-identical components, continuing with the most relevant conclusions of the work in the last Sect. 6.

2 Mathematical Modelling of the Problem

We consider a system composed of m monitored components and n non-monitored components. The assumptions that we made on the model are the following:

- The m monitored components are subject to a continuous degradation which is modelled using a Gamma process with shape parameter $\alpha_i(t)$ and scale parameter $\beta_i(t)$ depending on time t. Gamma processes are well-known and they are really useful in maintenance [2, 18]. If we denote by $X_i(t)$ the deterioration level of component i at time t, for $i = 1, \ldots, m$, and considering $\alpha_i(t) = \alpha_i t$ and $\beta_i(t) = \beta_i$, we have the density function of this process:

$$f_{(\alpha_i t, \beta_i)}(x) = \frac{{\beta_i}^{\alpha_i t}}{\Gamma(\alpha_i t)} x^{\alpha_i t - 1} \exp(-\beta_i x), \quad x \geq 0,$$

  where Γ is the well-known Gamma function.
- The n non-monitored components are subject to sudden failures [5]. We suppose that these failures follow a Poisson process, that is, the distribution of the time between failures of non-monitored components is an exponential distribution with parameter λ. Let Y be the time between failures, then the survival function of Y is given by $\bar{F}_Y(t) = \exp(-\lambda t)$.

The assumption of exponentially distributed times between inspections seems to be restrictive. Apart from the mathematical tractability of the Poisson process due to its memoryless property, the Poisson process is an excellent model for many real-world phenomena as a consequence of the *Palm-Khintchine Theorem* [17]. Supposing that we have a very large number of independent stochastic processes, where each separate process generates only rarely an event, the theorem states that the superposition of all these processes behaves approximately as a Poisson process.

That is the case of the degrading components in our model: the superposition of all of them, when $m \to \infty$ behaves as a Poisson process.

2.1 Maintenance Policy

The deterioration state of the system is inspected periodically, instead of having a continuous monitoring through sensors, which is more expensive. Apart from the inspection policy, a condition-based maintenance policy and an opportunistic maintenance policy are implemented. With condition-based maintenance [2, 3, 6], we perform maintenance actions depending on the system state. Adding opportunistic maintenance, this allows us to take advantage of the maintenance or inspection times of a component to check the state of the rest of the monitored components of the system.

For that, we set two different deterioration thresholds [14] in each monitored components: a corrective threshold L_i and a preventive threshold M_i, for $i \in 1, \ldots m$. At each inspection time, the deterioration level $X_i(t)$ of the monitored components is checked, for $i \in 1, \ldots m$. We have three situations:

1. If $X_i(t) < M_i$, then component i is left as it was.
2. If $M_i \leq X_i(t) < L_i$, then component i is preventively replaced by another one and $X_i(t)$ is now restored to zero.
3. If $X_i(t) \geq L_i$, then component i does not work and it is correctively maintained by replacing it.

This scheme is repeated for all $i \in I = \{1, 2, \ldots, m\}$. On the other hand, if a component reaches the preventive or corrective threshold, a signal is sent to the maintenance team, and it starts to repair the system after a certain time. With that, we can also check the system of the rest of the monitored components (with the same procedure that has been explained before) in these maintenance times, which is called opportunistic maintenance. We can see a representation of a Gamma process (which represents a simple system with only one component) in Fig. 1.

3 Expected Cost Definition

In this section we study the asymptotic behaviour of the expected cost of the system maintenance. A renewal cycle is the time between two maintenance actions that restored the system to the initial state, that is, to *as good as new* state. So that, we can treat the system as a Markov chain with a space of recurring states [8].

Let $\mathbb{E}[C^p(R)]$ and $\mathbb{E}[C^c(R)]$ be the costs due to the preventive and corrective maintenance of the monitored components in a renewal cycle R. Let $\mathbb{E}[C^{nm}(R)]$ be the expected cost due to the corrective replacements of the non-monitored

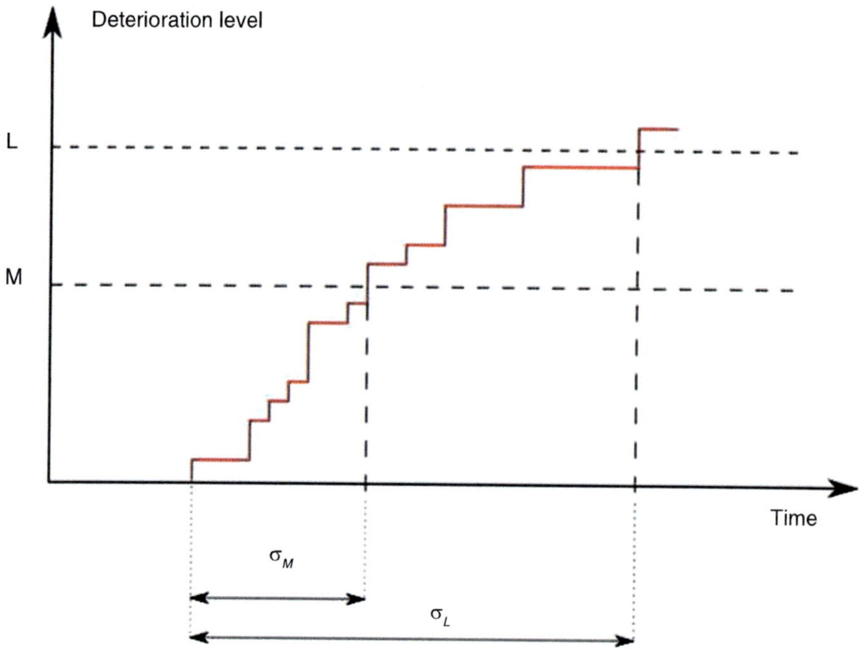

Fig. 1 Realization of a gamma degradation process with preventive threshold M and corrective threshold L. Random variables σ_M and σ_L denote the instant of time in which the preventive and corrective threshold are reached, respectively

components in a renewal cycle and $\mathbb{E}[C^i(R))]$ corresponds to the expected cost due to inspections.

Assigning a cost for each maintenance task, the long-run expected cost C_∞ can be obtained using a well-known result in renewal processes theory.

Theorem 1 (Renewal-Reward Theorem) *For a positive recurrent renewal process in which a reward R_j is earned during cycle length X_j and such that $\{(X_j, R_j) : j \geq 1\}$ is i.i.d. with $\mathbb{E}\left[|R_j|\right] < \infty$, the long run rate at which rewards are earned is given by:*

$$\lim_{t\to\infty} \frac{R(t)}{t} = \frac{\mathbb{E}[R]}{\mathbb{E}[X]}, \qquad \text{with probability 1.}$$

In other words, the rate at which rewards are earned is equal to the expected reward over a cycle divided by an expected cycle length. Moreover,

$$\lim_{t\to\infty} \frac{\mathbb{E}[R(t)]}{t} = \frac{\mathbb{E}[R]}{\mathbb{E}[X]}.$$

For $t \in (0, \infty)$, the average cost on the interval $[0, t]$ is $\frac{C(t)}{t}$, where $C(t)$ is the cost until time t. Using Theorem 1, the long-run expected cost is expressed as:

$$C_\infty = \lim_{t \longrightarrow \infty} \frac{\mathbb{E}[C(t)]}{t} = \frac{\mathbb{E}[C(O)]}{\mathbb{E}[O]},$$

where O is the length of a cycle, that is, in our case, the time to the next maintenance action. Finally, the expected cost rate is developed as:

$$C_\infty = \frac{\mathbb{E}[C^p(O)]}{\mathbb{E}[O]} + \frac{\mathbb{E}[C^c(O)]}{\mathbb{E}[O]} + \frac{\mathbb{E}[C^{nm}(O)]}{\mathbb{E}[O]} + \frac{\mathbb{E}[C^i(O)]}{\mathbb{E}[O]}. \tag{1}$$

4 Optimization Algorithms

The objective of the work is to find the optimal values for the time between inspections and the preventive threshold that minimize the expected cost. It is rather tricky due to the stochastic nature of the problem. For that, we are going to implement different optimization algorithms: an exhaustive search algorithm combined with different meta-heuristics algorithms (genetic algorithms, ant colony algorithms and pattern search algorithm).

4.1 Local Search

It is often successful to initially explore the solutions of the problem to get a rough idea of where the local maximum or minimum of the problem lies. At each iteration of the search algorithm, the condition is refined to change the current solution to a better one if necessary. As a first approximation of the solution of the problem it is a good method. For this local search, we define a search set for the time between inspections T and the preventive threshold M. For each combination of each pair of points, we calculate the minimum expected cost. At each iteration, we obtain the indices of the pair of points where we reach the minimum value, and perform a translation to a neighboring value in the initial search space (increasing a unit the indices and multiplying by an increment of 0.05 units), where we continue repeating the process. We finish when there is no improvement of the solution.

Combined with meta-heuristic techniques it is very useful to improve the solutions in an adequate computational time: initially a local search of the given problem is performed and then a meta-heuristic appropriate to the type of problem is applied.

4.2 Meta-Heuristic Algorithms

The problem studied is, in general, difficult to solve. Consequently, heuristic algorithms can be helpful. There optimization algorithms work very well for a certain type of problem, but provide bad solutions in other case [7].

In general, these techniques have certain stopping criteria. The most common are taking into account the mean fitness of the individuals or the fitness of the best individuals, and stop the searching when this value reaches a convergence value. In the case of the genetic algorithm, another usual stopping criteria is when the assigned number of generations is reached.

4.2.1 Genetic Algorithms

They are based on the evolution theory and natural selection. A genetic algorithm starts with a set (also called population) of N solutions (or individuals), which evolve according to some rules applied in the following iterations of the algorithm. Individuals are selected from the population and recombined, producing offspring which will be the next generation. They were first applied by Zio, Marseguerra and Compare to this problem in [7, 11, 12].

- **Initialization:** at time $t = 1$, we generate N random solutions. In this specific case of a maintenance optimization problem, the solutions is usually a vector of variables with the interval time between inspections, thresholds values, ... so they are of the form $(T_j, M_j),\ j = 1, \ldots, m$.
- **Fitness evaluation:** for the set of sampled parameters, the cost is evaluated. These values are collected and order to select the *best individuals* for the next iteration. A greater probability is assigned to the individuals that minimize the cost function.
- **Crossover:** an offspring is created by cutting the selected individuals into two parts and exchanging their parts. That is, given $S_1 = (T_1, M_1)$ and $S_2 = (T_2, M_2)$, the new individuals considered are $\bar{S_1} = (T_1, M_2)$ and $\bar{S_2} = (T_2, M_1)$.
- **Mutation:** it alters randomly some elements of a solution, always with a small probability. In this case, we randomly change a solution $S_i = (T_i, M_i)$ by replacing it with a nearby value, simulated by a uniform distribution.

Genetic Algorithm Pseudo-Code

```
Generate initial population randomly
    while i < max(iterations) / stop condition is not fulfilled
        do
            Fitness calculation of each individual
            Select the best individuals
            Perform crossover with a probability p
            Perform mutation with a small probability q
            Select the new individuals for the population
    end while
```

4.2.2 Pattern Search

Pattern search is a direct search used for optimizing functions that are not continuous or differentiable (also known as black-box search or derivative-free search).

At each iteration of the algorithm, it moves to the nearest point which best minimizes its objective function. This procedure is repeated until the accuracy is achieved, or the number of iterations (stopping criteria) of the algorithm is reached.

Pattern Search Algorithm Pseudo-Code

```
Initiate in a point p
 while i < max(iterations) / stop condition is not fulfilled
   do
     Evaluate the nearest neighbour of the point p
       if there is a best solution
          Update the current solution to the best neighbour
       else
          Move to the next point
       end if
 end while
```

4.2.3 Ant Colony Algorithm

Ant colony algorithm is useful in combinatorics and discrete problems with graphs. Samrout applied it in preventive maintenance optimization [16]. It is based on the behaviour of ants while they are seeking a path in the graph between their colony and a source of food. Ants move randomly and they discover the shortest path via pheromone trails. When more pheromone is deposited on path, it increases the probability of that path to be followed.

In the initialization phase, a set of edges and nodes is defined. Nodes are points enclosed in the search space $[0, T_{max}] \times [0, L_i]$, for $i \in 1, \ldots m$ and the edges are the connections between these nodes.

The probability that ant k chooses node j starting from node i is:

$$p(i, j) = \frac{\tau_{ij}(t)}{\sum_{k \in \text{Nodes}} \tau_{ik}(t)},$$

where $\tau_{ij}(t)$ is the amount of pheromone existing for edge I_{ij}.

Ant Colony Algorithm Pseudo-Code

```
Generate initial population of n ants
Initialize the pheromone path and the parameters
  while i < max(iterations) / stop condition is not fulfilled
      for each ant do
          Generate a solution using some transition rules
          Update the pheromone
      end for
    Refine with a local search
```

```
    Choose the best ant
    Update the best global solution
end while
```

5 Case Study

The study of the previous maintenance model is completed with the search of the optimal values for the time between inspections, T_{opt} and the preventive threshold M_{opt} that minimize the total expected cost in Eq. (1). This search is performed using the software MATLAB. A combined method with Monte-Carlo simulation and meta-heuristics techniques is applied to perform the optimization [11, 12].

5.1 *Identical Degrading Components*

We study a system with m identical degrading components. So, $\alpha_i = \alpha$, $\beta_i = \beta$ and $L_i = L$ for all $i = 1, \ldots, m$. The following parameters and sequence of costs are imposed in the computation process:

- Cost of a corrective maintenance: 80 monetary units.
- Cost of a preventive maintenance: 30 monetary units.
- Cost due to periodic inspections: 30 monetary units
- Cost due to non-monitored components: 80 monetary units.
- Corrective threshold L: 6.
- Shape parameter α of the gamma process: 1.25.
- Scale parameter β of the gamma process: 0.5.

The objective is to find the optimal values T_{opt} and M_{opt} for the time between inspections and the preventive threshold M, respectively, that minimize the total expected cost of the system maintenance. The search space for the optimal preventive threshold M is $[0, L]$, and for the time T between inspections is $[0, T_{max}]$, with $T_{max} = 15$ time units, due to it is a logical interval value if we take into account the rest of the model parameters. In Table 1, the optimal values for M are represented, obtained with three different meta-heuristics. Ant Colony (ACO) works worse if we compare it to the results obtained with the others. The ones obtained with the Genetic Algorithm and the Pattern Search Algorithm are more similar. CPU time is expressed in minutes.

The optimal values T_{opt} for the time between inspection T are shown in Table 2. In both cases, Genetic Algorithm is the most efficient in relation to computational time and the optimal expected cost obtained, which is shown in Table 3. Due to this fact, the best algorithm to optimize our problem is the Genetic Algorithm, and it has been implemented to study the total expected cost when increasing the number of components.

Table 1 Optimization of preventive threshold M with Genetic Algorithm, Ant Colony Algorithm and Pattern Search Algorithm

Components	M_{opt} GA	CPU time	M_{opt} ACO	CPU time	M_{opt} PS	CPU time
1	2.87	2.25	2.27	2.58	2.69	1.45
2	2.90	2.10	2.37	4.27	2.98	3.10
3	3.17	3.43	2.50	6.23	3.40	4.46
4	3.29	3.57	2.44	7.98	2.98	5.76
5	3.27	4.53	2.51	8.34	2.97	6.64
6	3.30	4.67	2.65	9.89	3.02	8.32
7	3.48	5.28	2.61	10.67	3.14	8.56
8	4.05	5.96	3.30	11.23	3.83	9.02
9	4.32	6.47	4.27	11.65	3.97	9.87
10	4.95	8.21	5.04	15.32	4.56	11.47

Table 2 Optimization of inspection time T with Genetic Algorithm, Ant Colony and Pattern Search Algorithm

Components	T_{opt} GA	CPU time	T_{opt} ACO	CPU time	T_{opt} PS	CPU time
1	2.90	2.51	2.79	1.98	2.69	2.87
2	2.34	2.66	2.59	2.78	2.98	3.46
3	2.20	3.24	1.80	3.46	3.4	5.11
4	2.20	5.01	1.77	5.00	2.98	5.56
5	1.68	5.32	1.79	6.38	2.97	6.63
6	1.69	6.18	1.48	7.89	3.02	7.42
7	1.30	6.74	1.30	8.05	3.10	8.44
8	1.22	7.23	1.43	9.36	2.48	9.24
9	1.15	8.19	1.19	10.55	1.67	10.22
10	1.00	10.04	0.95	11.87	1.24	10.61

Table 3 Expected cost C_∞ calculated with local search and Genetic Algorithm

Components	Expected cost (C_∞)
1	4.8970
2	6.8499
3	8.6851
4	10.2468
5	11.6907
6	12.9925
7	14.1216
8	15.0826
9	16.5598
10	16.7657

5.2 *Non-Identical Degrading Components*

The system is considered to be made up of non-identical degrading components, with different scale parameter in the gamma process that defines their deterioration. Therefore, the preventive threshold will be different for each of the m monitored components.

We maintain the same maintenance costs as in the case of identical components as in Sect. 5.1. The corrective threshold remains fixed and the same for all components. However, the shape parameter of the gamma process will now vary, as will the preventive threshold, which will be different for each of the components as they have different degradation processes.

Therefore, the shape parameter increases by 0.1 units as a component is added to the system, as follows

$$\alpha_1 = 1.2, \qquad \alpha_i = \alpha_{i-1} + 0.1, \qquad \text{for } i \in 1, \ldots, m$$

The same search space for T_{opt} and M_{opt} as in the previous section is used, and the objective is to optimize these parameters in order to obtain the minimum expected cost. With different components, we are calculating initial points in Table 4, using the local search and after that, the Genetic Algorithm is applied to look for the best solution. The results are shown in Table 5.

Table 4 Initial points obtained with Local Search Algorithm for different components

Components	T_0	M_0
1	4.65	2.54
2	4.39	2.95
3	4.28	3.17
4	3.13	3.37
5	2.47	3.78
6	2.29	3.93
7	1.89	4.18
8	1.64	4.37
9	1.53	4.65
10	1.48	4.70

Table 5 Optimal values T_{opt}, M_{opt} and C_∞ obtained with Genetic Algorithm

Components	T_{opt}	M_{opt}	C_∞
1	6.29	2.37	5.24
2	5.99	(2.68, 2.01)	6.32
3	3.93	(3.39, 3.29, 3.08)	7.85
4	2.79	(4.31, 3.95, 3.49, 3.23)	9.96
5	2.20	(3.93, 3.86, 3.80, 3.72, 3.66)	10.94
6	2.34	(3.95, 4.25, 4.62, 4.28, 4.47, 4.12)	12.45
7	2.57	(4.58, 4.75, 4.66, 4.98, 4.87, 5.01, 4.57)	14.52
8	2.45	(4.52, 4.17, 4.36, 4.40, 4.84, 4.10, 4.74, 4.65)	16.16
9	2.18	(4.14, 4.38, 4.45, 4.36, 4.02, 4.11, 4.54, 4.32, 4.15)	19.73
10	2.37	(3.85, 4.15, 4.26, 4.02, 3.48, 3.87, 3.96, 4.00, 3.88, 3.98)	22.18

6 Conclusions

The main novelty of this work is the use of heuristics to find the best maintenance strategy, which allows us to save cost. Previous work has focused on modeling the problem from a probabilistic and analytical point of view [5, 6]. Another point in favor is to perform a maintenance policy at the system level, considering the components as a whole when performing maintenance based on the system condition, instead of doing it component by component, as in [4, 9] or [19].

The optimal cost, as well as the optimal preventive thresholds and time between inspections for a system consisting of identical and non-identical components have been obtained. The computational time is reduced if we initially perform a local search for the parameters T and M. The most efficient algorithm is the genetic algorithm. It is observed that the expected cost increases as more components are added. In the case of identical components, the optimal value for the preventive threshold is increasing, while the optimal time between inspections is decreasing as more components are added. Something similar occurs with non-identical components, although M_{opt} and T_{opt} stabilize as the number of components $m \to \infty$.

In future work, it would be interesting to consider a dependency between degrading and non-degrading components. The assumption of exponential distribution between failure times thanks to the Palm-Khintchine theorem will allow us to greatly simplify this objective. Another interesting point would be the improvement of the algorithms already proposed for the resolution of the problem.

Acknowledgments This research was supported by Junta de Extremadura, Spain (Project GR18108) and European Union (European Regional Development Funds).

The authors want to thank the anonymous reviewers for their useful comments and time.

References

1. Ab-Samat, H., Kamaruddin, S.: Opportunistic maintenance (OM) as a new advancement in maintenance approaches. J. Qual. Maint. Eng. **20**(2), 98–121 (2014)
2. Asmussen, S.: Applied probability and queues. Applications of Mathematics vol. 51, 2nd edn. Springer, New York (2003)
3. Bertoin, J.: Lévy Processes. Cambridge University Press (1996)
4. Castanier, B., Grall, A., Bérenguer, C.: A condition-based maintenance policy with non-periodic inspections for a two-unit series system. Reliab. Eng. Syst. Saf. **87**(1), 109–120 (2005)
5. Castro, I.T., Basten, R.J.I., van Houtum, G.J.: Maintenance cost evaluation for heterogeneous complex systems under continuous monitoring. Reliab. Eng. Syst. Saf., 106745 (2020)
6. Caballé, N., Castro, I.T.: Analysis of the reliability and the maintenance cost for finite life cycle systems subject to degradation and shocks. Appl. Math. Modell. **52**, 731–746 (2017)
7. Compare, M., Martini, F., Zio, E.: Genetic algorithms for condition-based maintenance optimization under uncertainty. Eur. J. Oper. Res. **244**(2), 611–623 (2015)
8. de Jonge, B., Klingenberg, W., Teunter, R., Tinga, T.: Reducing costs by clustering maintenance activities for multiple critical units. Reliab. Eng. Syst. Saf. **145**, 93–103 (2016)
9. Golmakani, H.R., Moakedi, H.: Periodic inspection optimization model for a two-component repairable system with failure interaction. Comput. Ind. Eng. **63**(3), 540–545 (2012)
10. Grall, A., Dieulle, L., Bérenguer, C., Roussignol, M.: Continuous-time predictive-maintenance scheduling for a deteriorating system. IEEE Trans. Reliab. **51**(2), 141–150 (2002)
11. Marseguerra, M., Zio, E.: Condition-based maintenance optimization by means of genetic algorithms and Monte Carlo simulation. Reliab. Eng. Syst. Saf. **77**(2), 151–165 (2002)
12. Marseguerra, M., Zio, E.: Optimizing maintenance and repair policies via a combination of genetic algorithms and Monte Carlo simulation. Reliab. Eng. Syst. Saf. **68**(1), 69–83 (2000)
13. Minou, C.A., Keizer, O., Flapper, S.D.P., Teunter, R.H.: Condition-based maintenance policies for systems with multiple dependent components: A review. Eur. J. Oper. Res. **261**(2), 405–420 (2017)
14. Poppe, J., Boute, R.N., Lambretch, M.R.: A hybrid condition-based maintenance policy for continuously monitored components with two degradation thresholds. Eur. J. Oper. Res. **268**(2), 515–532 (2018)
15. Rasmekomen, N., Parlikad, A.K.: Condition-based maintenance of multi-component systems with degradation state-rate interactions. Reliab. Eng. Syst. Saf. **148**, 1–10 (2010)
16. Samrout, M., Kouta, R., Yalaoui, F., Châtelet, E.: Parameter's setting of the ant colony algorithm applied in preventive maintenance optimization. J. Intell. Manuf. **18**, 663–677 (2007)
17. Tijms, H.C.: A First Course in Stochastic Models. Wiley (2018)
18. van Noortwijk, J.M.: A survey of the application of gamma processes in maintenance. Reliab. Eng. Syst. Saf. **94**(1), 2–21 (2009)
19. Yang, L., Zhao, Y., Ma, X., Qiu, Q.: An optimal inspection and replacement policy for a two-unit system. Proc. Inst. Mech. Eng. 0 J. Risk Reliab. **232**(6), 766–776 (2018)

Metal Additive Manufacturing: Nesting vs. Scheduling

Ibrahim Kucukkoc

Abstract As a new and emerging technology, additive manufacturing (AM) is considered to be the future of manufacturing. In comparison to traditional machining techniques, parts are produced through a layer-by-layer production process in AM. It provides both design flexibility and strength as well as lightness and low prototyping costs. Selective laser melting (SLM) is a popular AM technology used in fabricating metal components. Although the SLM technology is a high-cost process at first glance, it can be compensated with efficient planning and scheduling systems. This chapter aims to investigate the relationship between nesting and scheduling when planning and scheduling SLM machines. Numerical examples are conducted to show that optimal nesting does not guarantee optimal scheduling. It is concluded that the nesting and scheduling problems must be considered simultaneously to ensure feasibility.

Keywords Additive manufacturing · 3D printing · Selective laser melting · Scheduling · Nesting · Planning · Metaheuristics

1 Introduction

Additive manufacturing (also known as 3D printing) techniques have appeared in 1980s and evolved significantly since then. Among those [1], fused deposition modelling, laminated object manufacturing, stereo lithography and selective laser sintering are used for non-metal products; selective laser melting (SLM), electron beam melting and laser engineered net shaping are utilized for fabricating metal parts. Ngo et al. [2] presented a review on materials, methods, applications and challenges in AM.

I. Kucukkoc (✉)
Department of Industrial Engineering, Balikesir University, Balikesir, Turkey
e-mail: ikucukkoc@balikesir.edu.tr

A. Masone et al. (eds.), *Optimization and Data Science: Trends and Applications*, AIRO Springer Series 6, https://doi.org/10.1007/978-3-030-86286-2_13

Thanks to its layer-by-layer production process, additive manufacturing (AM) processes provide many significant advantages over traditional subtractive techniques. Some of these advantages are strength to weight ratio, design flexibility, high accuracy and resource efficiency [3].

With enormous advancements gained in the materials and manufacturing technology, some particular techniques (such as SLM) have shifted from prototyping to direct digital manufacturing. This yielded a revolution in several industries such as defense, aerospace, automotive and healthcare industries. In comparison to its competitors in metal AM domain, SLM has become the dominant application of metal AM processes thanks to its high accuracy and performance. Many leading aerospace and aeronautic companies, including NASA, Airbus and Boeing, utilize additive manufacturing to produce a large percentage of components (including the critical ones such as fuel pumps and nozzles) as a single part instead of assembly. Ford and Despeisse [4] summarized the advantages and challenges of AM and provided a discussion regarding the implications on sustainability.

This chapter addresses the multiple SLM machines scheduling problem considering nesting constraints (i.e. 2D allocation of parts with no overlap). The main objective of the chapter is to numerically show the necessity of handling both nesting and scheduling problems together. The numerical tests also show that optimal nesting does not mean optimal scheduling in the AM machine scheduling problem.

The rest of the chapter is organized as follows. Sect. 2 provides the review of the literature and Sect. 3 presents a formal description of the problem studied. Section 4 presents the details of the solution method, adapted from Kucukkoc et al. [5] and Sect. 5 represents the results of the numerical examples. The chapter is concluded in Sect. 6, followed by some future research directions.

2 Literature Review

There has been an extensive amount of research on the AM technology itself. A recent review on the advancement of materials and methods involved in AM has been presented by Ngo et al. [2]. However, the topic covered in the current work is more about the planning of AM resources, i.e. machines. This is an emerging field as most of the research on AM has focused on the process and its applications in different industries rather than the efficient use of AM machines (see, for example [6–9]). Among others, Li et al. [10] investigated the influence of scan length on fabricating thin-walled components in SLM. Camacho et al. [11] presented a review of AM applications in the construction industry. Li et al. [22] applied response surface methodology to optimize the process parameters of SLM with the aim of minimizing surface roughness. Frazier [12] provided a review on SLM.

Regarding the production planning of SLM machines, the number of researches is quite limited but increasing rapidly. The preliminary researches on planning and scheduling of SLM machines belong to Kucukkoc et al. [5] and Li et al. [13]. The problem was introduced in Kucukkoc et al. [5], in which a mathematical model was proposed to maximize resource utilization considering part delivery times. Li et al. [13] proposed a mixed-integer linear programming (MILP) model and two heuristics, called best-fit and adapted best-fit, to minimize the average production cost per material volume. Kucukkoc et al. [14] developed a genetic algorithm (GA) approach to minimise maximum lateness in multiple SLM machines. Chergui et al. [15] proposed a heuristic approach for the parallel identical AM machines scheduling and nesting problem to minimize maximum lateness. Zhang et al. [16] focussed on the two-dimensional placement optimization of AM parts providing a two-step strategy. Kucukkoc [3] proposed MILP models to minimise makespan in (a) single, (b) parallel related, and (c) parallel unrelated AM machine scheduling problems. Che et al. [17] considered orientation selection and two-dimensional nesting problems simultaneously for unrelated parallel AM machine scheduling problem. Aloui and Hadj-Hamou [18] proposed a heuristic for AM machine scheduling problem under technological constraints. Oh et al. [19] reviewed the studies on nesting and scheduling problems for AM.

As investigated in this survey, to the best knowledge of the authors, there is no research which discusses how nesting affects scheduling (and vice versa) when planning AM machines. The contribution of this chapter is based on numerical problems to show that better solutions in terms of nesting objectives are not the best solutions in terms of scheduling objectives in AM machine scheduling problems.

3 Problem Statement

Multiple AM machine scheduling problem is the problem of determining [3]:

(i) The allocation of heterogenous parts into jobs and
(ii) The utilization order of jobs on different AM machines with different characteristics

To optimize one or more performance criterion such as makespan, total or average cost, and total or maximum tardiness. Figure 1 represents a schematic representation of multiple AM machine scheduling problem.

Parts, jobs and machines are denoted by i, j and m, respectively. I is the set of parts, J is the set of jobs and M is the set of machines. Each part has a height (h_i), area (a_i), volume (v_i), width (w_i), length (l_i) and release date (r_i). Machines may have different specifications represented as follows:

VT_m Time spent to form per unit volume of material by machine m
HT_m Time spent for powder-layering by machine m, which is repeated in each layer in accordance with the highest part in the job
SET_m Set-up time needed for initializing and cleaning machine m
MA_m The production area of the machine m's building platform
MW_m The width of the machine m's building platform
ML_m The length of the machine m's building platform

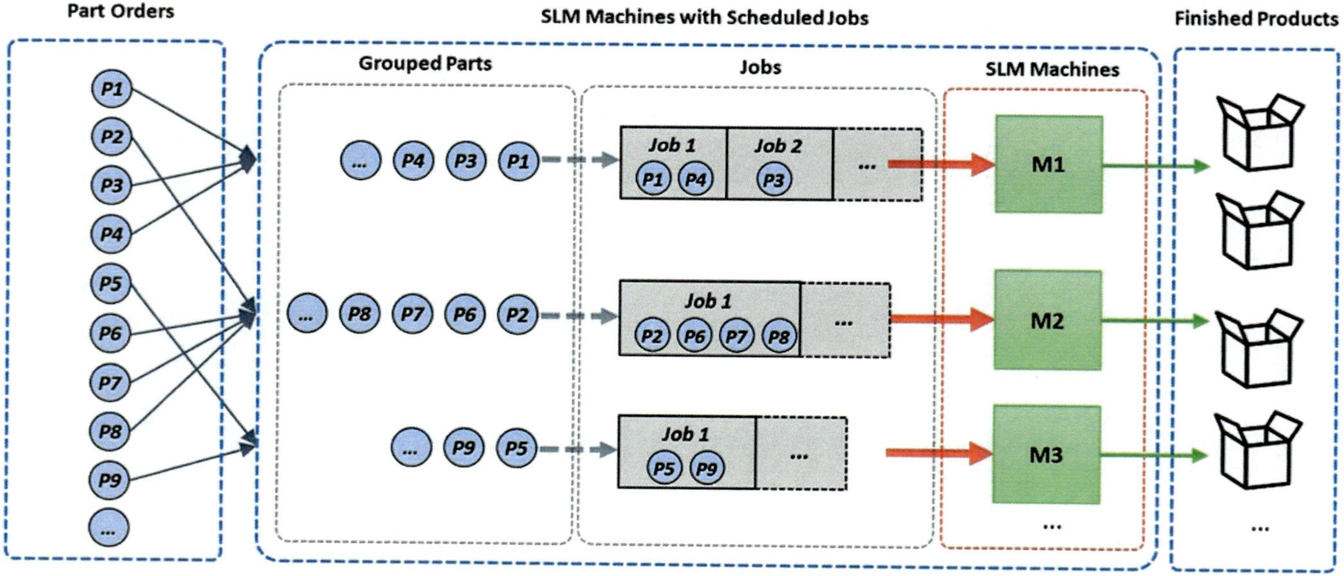

Fig. 1 A schematic representation of multiple AM machine scheduling problem, adapted from Li et al. [13]

As indicated by Kucukkoc [3] the AM machine scheduling problem is different from the batch scheduling problem in several aspects. One of the most peculiar features of the AM machine scheduling problem is that the production time of a job is calculated via a function based on some characteristics of parts produced in that job (whereas in a classical batch scheduling problem the job production time is known in advance). That makes it impossible to calculate the processing time of any individual part prior to scheduling (i.e. deciding on the part content of the job on a specific machine). Also, different combinations of parts in jobs lead to different performance measures and costs in the AM machine scheduling problem. Furthermore, increasing the utilization of batch scheduling machines is preferred as it generally has a positive impact on minimizing the makespan or cost. However, this is not always the case for AM machine scheduling problems. This chapter addresses to this main issue. The AM machine scheduling problem includes both nesting and scheduling problems each of which is NP-hard itself.

Using the notation provided, the production time (PT_{mj}) and the completion time (C_{mj}) for job j on machine m can be calculated using Eqs. (1) and (2), respectively [3]. C_{m0} is assumed to be 0 (see Eq. (3)).

$$PT_{mj} = SET_m \cdot Z_{mj} + VT_m \cdot \sum_{i \in I} v_i \cdot X_{mji} + HT_m \cdot \max_{i \in I} \left\{ h_i \cdot X_{mji} \right\}; \qquad \forall m \in M, j \in J. \tag{1}$$

$$C_{mj} \geq C_{m(j-1)} + PT_{mj}; \qquad \forall m \in M, j \in J. \tag{2}$$

$$C_{m0} = 0; \qquad \forall m \in M. \tag{3}$$

PT_{mj} is constituted of three terms: (a) set-up time, (b) material release time, and (c) powder-layering time. To include the set-up, a binary decision variable (Z_{mj}) is utilized. It gets the value of 1 if job j is utilized on machine m, 0 otherwise. The material release time is related to the total volume of parts included in job j on machine m. X_{mji} is a binary variable which gets 1 if part i is assigned to job j on machine m, 0 otherwise. Lastly, powder-layering time is related to the highest part included in a job. This indicates that grouping parts with similar heights has a potential to minimize job production time. The ultimate goal is to minimise makespan and the assumptions can be summarized as follows:

- Part details are known exactly.
- There is more than one machine with the same or different characteristics.
- At least one of the machines has the capability of fabricating the part with the maximum width/length/height.
- Any number of machines can be operated simultaneously.
- There is no restriction between any two parts to be produced in the same job.
- The building orientation of parts is known.
- All parts are made of the same material.
- No part can be removed until the job is completed.

The following section presents the solution method used for numerical problems.

4 Solution Methodology

GA has been utilized successfully in solving various scheduling problems. According to Lee [20], over 55% of the researches on operations management belong to operations planning and control. Moreover, 57% of those studies address scheduling problems. Therefore, this chapter applies GA utilized in Kucukkoc et al. [21] for the solution of complex scheduling problem with the aim of optimizing two different objectives under two scenarios, i.e. maximizing area utilization and minimizing makespan. The general outline of the GA is presented in Fig. 2.

This is a classical GA which applies tournament selection to determine parents which will undergo genetic operators (crossover and mutation). The chromosomes are permutation-coded, where each gene represents a part number. So, the length of the chromosome is equal to the number of parts to be scheduled. A single-point crossover is applied ensuring that no gene duplicates on the chromosome. Mutation is applied in two ways using either an insert function or a swap function. The new generation is formed replacing the worst chromosomes in the population. The algorithm terminates if there is no improvement within a certain number of iterations (*MaxIt*) determined priorly.

The critical point of the utilized algorithm is the *Nesting()* function employed in decoding procedure given in Fig. 3. This function is employed to check whether the

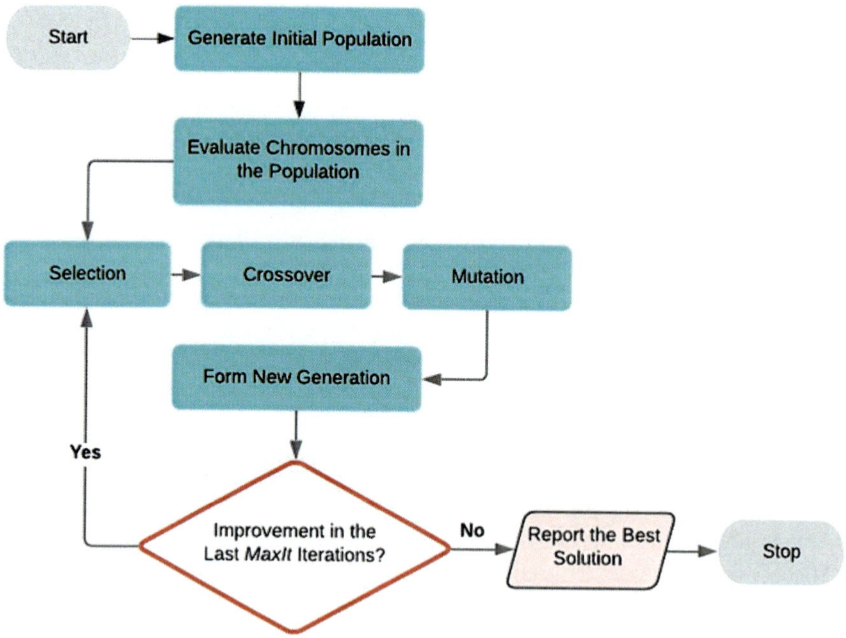

Fig. 2 The main outline of the GA

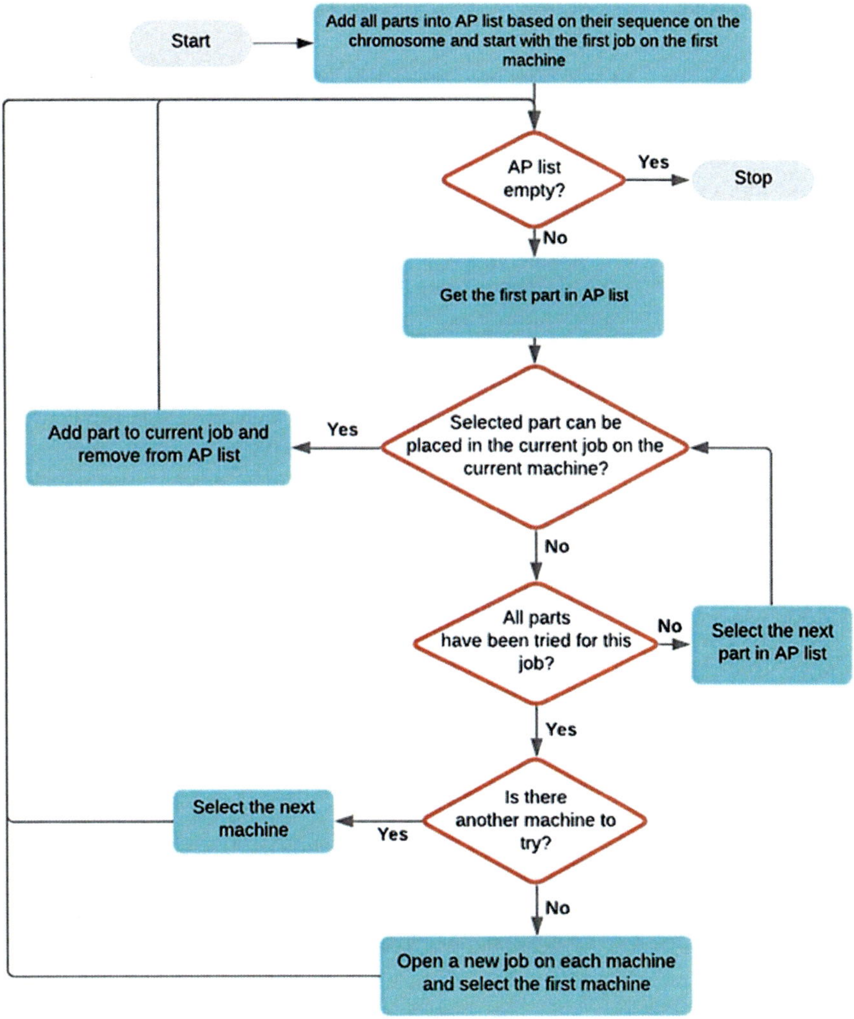

Fig. 3 Decoding procedure applied to evaluate chromosomes (AP denotes available parts)

selected part can be placed in the current job on the current machine. It ensures that the parts assigned to a job on a specific machine can be placed on the corresponding machine's platform with no overlap. More details on the search mechanisms of the algorithm can be found in Kucukkoc et al. [21].

5 Numerical Examples

The algorithm has been coded in JAVA and run on Intel® Core™ i7-6700HQ CPU @2.60 GHz with 16 GB of RAM. Each problem has been solved three times and the best result is reported. Two different problems are constituted by different number of parts and machines. Two types of machines are considered and their details are given in Table 1. Parts data have been provided in Appendix (retrieved from [21]). The population size, crossover rate and mutation rate are 30, 0.6 and 0.1, respectively. The algorithm was terminated when there is no improvement within 3000 iterations. The values of these parameters have been determined through some preliminary tests.

The first problem aims to schedule a total of 18 parts on two machines (one from type M1 and one from type M2). The parts range considered in this problem is P1-P18 (see Appendix). The total area of the parts is 2616.6 cm^2.

Figure 4 shows the solution obtained when the problem has been solved with the aim of minimizing makespan. As can be seen from this, the makespan is 507.33 h and the total area allocated is 4375 cm^2. Figure 5 presents another solution, which requires one fewer (six) jobs, for the same problem. The makespan for this solution is 627.07 h and the total area utilized is 3750 cm^2. When the two alternative solutions are compared, it is obvious that the area utilization of the second solution is higher (69.77% versus 59.80%). However, the makespan of the first solution is much lower than that of the second solution.

Table 1 Machine details

Machine ID	Height (cm)	Width (cm)	Length (cm)	Area (cm^2)	Set-up Time (h)	HT (h/cm)	VT (h/cm^3)
M1	32.5	25	25	625	2.00	1	0.030864
M2	32.5	25	25	625	1.00	1	0.030864

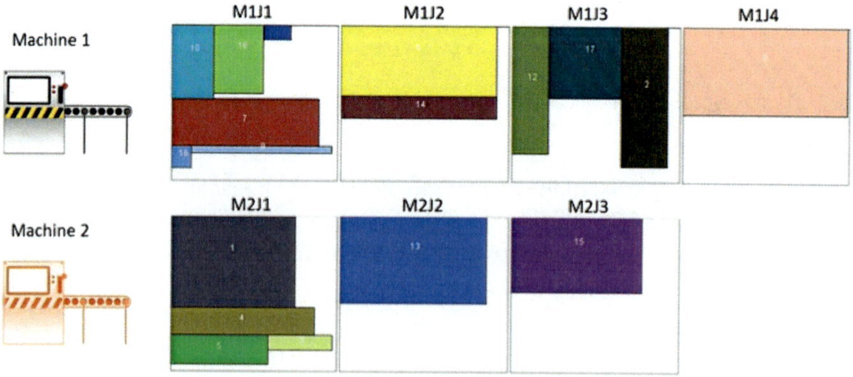

Fig. 4 The first alternative solution for problem-1 (507.33 h, 4375 cm^2)

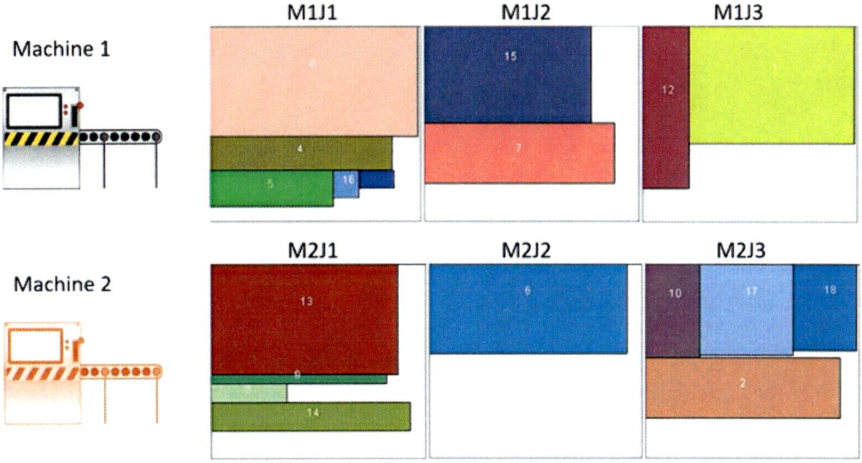

Fig. 5 The second alternative solution for problem-1 (627.07 h, 3750 cm^2)

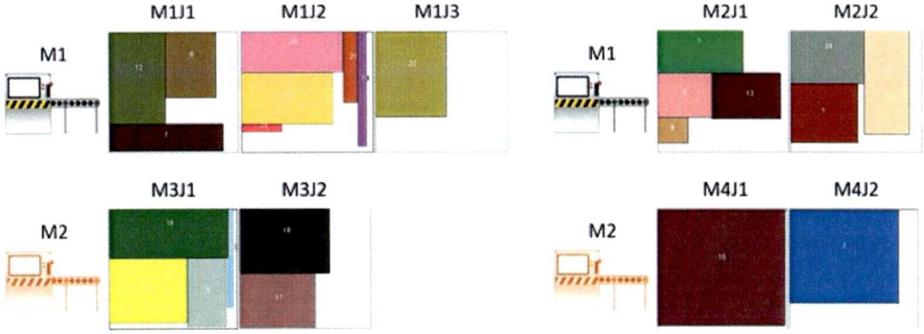

Fig. 6 The first alternative solution for problem-2 (653.31 h, 5625 cm^2)

The second problem involves 24 parts (ranging between P17 and P40, see Appendix) and 4 machines (2xM1 and 2xM2). Note that in the algorithm the parts have been re-numerated starting from 1. The total area of the parts is 4066.6 cm^2. The two alternative solutions obtained for this problem using GA are given in Figs. 6 and 7. A total of 9 jobs have been utilized in the first solution, with performance measures of 653.31 h makespan and 5625 cm^2 total area utilized. The second alternative solution requires two more (i.e., 11) jobs and the makespan is 636.06 h while the total utilized area is 6875 cm^2. As clearly seen from these results, the second solution requires a greater number of jobs and so total area (22% larger than the first solution). However, the makespan of the second solution is lower than the makespan of the first solution.

The results of these two problems indicate that the minimization of the total number of jobs (or total area required) does not always correspond to the minimization of the makespan. Therefore, nesting problem needs to be considered in

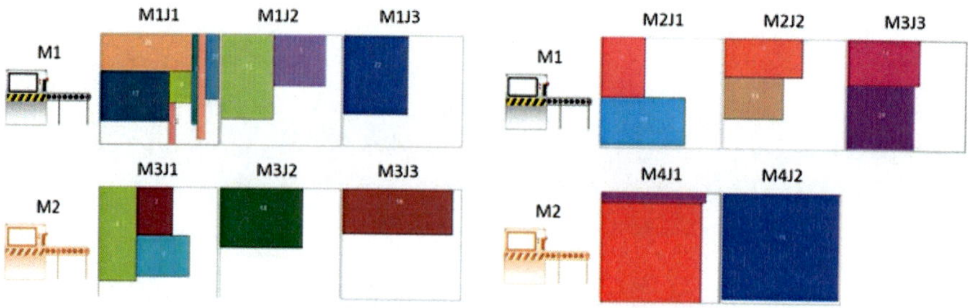

Fig. 7 The second alternative solution for problem-2 (636.06 h, 6875 cm^2)

the AM machine scheduling problem as a constraint to ensure feasibility. However, considering it as an ultimate goal may yield poor solutions in terms of the makespan.

6 Conclusion and Future Research

The multiple AM machine scheduling problem with two-dimensional nesting constraints has been studied in this chapter. A GA algorithm has been utilized to solve numerical examples and represent detailed nesting solutions. The aim was to investigate the relationship between the total area utilized and the makespan (i.e. the completion time of the latest part in the system). The results of the numerical examples indicated that there is no direct relationship between minimizing the number of jobs and minimization of the makespan. In both problems provided, the solutions with fewer number of jobs (and so area utilized) took longer to have all parts completed. One limitation of this study is that more experiments are needed to conclude that area utilization does not correspond to minimizing makespan. The author's ongoing work aim to design and conduct a comprehensive set of experimental tests for this aim. Developing new algorithms eliminating some assumptions of the problem can also be a research for future studies.

A.1 Appendix

Table A.1 Part details

Part	h_i	w_i	l_i	a_i	v_i	r_i
P1	16.7	18.8	16.0	300.8	1573.8	6.25
P2	8.8	6.7	22.8	152.8	421.3	7.35
P3	20.3	9.3	2.1	19.5	147.8	9.83
P4	7.4	21.6	3.9	84.2	285.2	20.95
P5	27.3	14.9	4.1	61.1	583.3	36.58
P6	25.8	23.2	12.9	299.3	3282.5	51.50
P7	14.5	22.2	6.7	148.7	1265.5	56.35
P8	3.5	24.6	15.3	376.4	723.3	69.90
P9	20.4	20.5	1.0	20.5	278.5	75.07
P10	23.3	6.8	13.4	91.1	1051.8	86.00
P11	26.3	4.0	5.0	20.0	201.9	93.03
P12	12.8	5.9	21.0	123.9	866.1	93.10
P13	28.4	21.8	15.3	333.5	7347.6	97.15
P14	10.9	23.2	3.2	74.2	333.6	97.25
P15	14.1	19.6	13.7	268.5	2956.0	101.75
P16	3.9	3.6	3.1	11.2	32.5	106.95
P17	3.2	10.6	13.1	138.9	265.6	107.58
P18	24.6	7.3	12.6	92.0	1387.4	115.55
P19	7.1	21.2	20.0	424.0	1086.0	128.37
P20	25.2	8.4	21.6	181.4	3559.2	132.32
P21	29.6	16.4	10.6	173.8	2902.9	134.15
P22	21.1	10.8	8.0	86.4	854.6	136.65
P23	25.2	22.0	4.9	107.8	1986.0	138.30
P24	28.2	9.5	14.9	141.6	2974.3	149.48
P25	22.9	6.3	4.6	29.0	408.2	164.08
P26	24.2	17.7	9.4	166.4	1271.4	167.45
P27	19.0	1.0	20.9	20.9	252.4	177.23
P28	26.6	11.2	19.7	220.6	3740.9	206.40
P29	18.5	12.5	8.2	102.5	991.9	219.62
P30	10.1	15.4	11.9	183.3	1377.6	228.55
P31	24.5	24.5	24.2	592.9	11,122.9	242.90
P32	30.6	23.0	11.9	273.7	5234.1	244.87
P33	3.1	14.7	9.9	145.5	274.6	245.08
P34	18.9	17.4	14.8	257.5	2628.4	248.18
P35	11.2	1.7	23.9	40.6	206.9	258.98
P36	4.6	19.5	10.4	202.8	639.5	268.42
P37	23.4	2.6	15.9	41.3	431.5	279.77
P38	7.5	13.7	18.5	253.5	968.2	286.25
P39	12.7	8.2	1.4	11.5	95.0	290.73
P40	22.7	14.3	12.5	178.8	2306.2	294.53

References

1. Attaran, M.: The rise of 3-D printing: the advantages of additive manufacturing over traditional manufacturing. Bus. Horiz. **60**(5), 677–688 (2017)
2. Ngo, T.D., Kashani, A., Imbalzano, G., Nguyen, K.T.Q., Hui, D.: Additive manufacturing (3D printing): a review of materials, methods, applications and challenges. Compos. Part B. **143**, 172–196 (2018)
3. Kucukkoc, I.: MILP models to minimise makespan in additive manufacturing machine scheduling problems. Comput. Oper. Res. **105**, 58–67 (2019)
4. Ford, S., Despeisse, M.: Additive manufacturing and sustainability: an exploratory study of the advantages and challenges. J. Clean. Prod. **137**, 1573–1587 (2016)
5. Kucukkoc, I., Li, Q., Zhang, D. Z.: Increasing the utilisation of additive manufacturing and 3D printing machines considering order delivery times. In: 19th International Working Seminar on Production Economics, February 22–26, Innsbruck, Austria (2016)
6. Cooper, D.E., Stanford, M., Kibble, K.A., Gibbons, G.J.: Additive manufacturing for product improvement at Red Bull Technology. Mater. Des. **41**, 226–230 (2012)
7. Javaid, M., Haleem, A.: Additive manufacturing applications in orthopaedics: a review. J. Clin. Orthop. Trauma. **9**(3), 202–206 (2018)
8. Khajavi, S.H., Partanen, J., Holmström, J.: Additive manufacturing in the spare parts supply chain. Comput. Ind. **65**(1), 50–63 (2014)
9. Li, Z., Zhang, D.Z., Dong, P., Kucukkoc, I., Peikang, B.: Incorporating draw constraint in the lightweight and self-supporting optimisation process for selective laser melting. Int. J. Adv. Manuf. Technol. **98**(1), 405–412 (2018)
10. Li, Z., Xu, R., Zhang, Z., Kucukkoc, I.: The influence of scan length on fabricating thin-walled components in selective laser melting. Int. J. Mach. Tools Manuf. **126**, 1–12 (2018)
11. Camacho, D.D., Clayton, P., O'Brien, W.J., Seepersad, C., Juenger, M., Ferron, R., Salamone, S.: Applications of additive manufacturing in the construction industry—a forward-looking review. Autom. Constr. **89**, 110–119 (2018)
12. Frazier, W.E.: Metal additive manufacturing: a review. J. Mater. Eng. Perform. **23**(6), 1917–1928 (2014)
13. Li, Q., Kucukkoc, I., Zhang, D.Z.: Production planning in additive manufacturing and 3D printing. Comput. Oper. Res. **83**, 157–172 (2017)
14. Kucukkoc, I., Li, Q., He, N., Zhang, D.: Scheduling of multiple additive manufacturing and 3D printing machines to minimise maximum lateness. In: 20th International Working Seminar on Production Economics, February 19–23, Innsbruck, Austria (2018)
15. Chergui, A., Hadj-Hamou, K., Vignat, F.: Production scheduling and nesting in additive manufacturing. Comput. Ind. Eng. **126**, 292–301 (2018)
16. Zhang, Y., Gupta, R.K., Bernard, A.: Two-dimensional placement optimization for multi-parts production in additive manufacturing. Robot. Comput. Integr. Manuf. **38**, 102–117 (2016)
17. Che, Y., Hu, K., Zhang, Z., Lim, A.: Machine scheduling with orientation selection and two-dimensional packing for additive manufacturing. Comput. Oper. Res. **130**, 105245 (2021)
18. Aloui, A., Hadj-Hamou, K.: A heuristic approach for a scheduling problem in additive manufacturing under technological constraints. Comput. Ind. Eng. **154**, 107115 (2021)
19. Oh, Y., Witherell, P., Lu, Y., Sprock, T.: Nesting and scheduling problems for additive manufacturing: a taxonomy and review. Addit. Manuf. **36**, 101492 (2020)
20. Lee, C.K.H.: A review of applications of genetic algorithms in operations management. Eng. Appl. Artif. Intell. **76**, 1–12 (2018)
21. Kucukkoc, I., Li, Z., Li, Q.: 2D nesting and scheduling in metal additive manufacturing. In: The International Conference of Production Research, ICPR—Americas 2020, 9–11 December 2020, Bahia Blanca, Argentina (2020)
22. Li, Z., Kucukkoc, I., Zhang, D.Z., Liu, F.: Optimising the process parameters of selective laser melting for the fabrication of Ti6Al4V alloy. Rapid Prototyp. J. **24**(1), 150–159 (2018)

System and Methods for Blockchain-Inspired Digital Game Asset Management

Gianluca Ragnoni

Abstract In 2017, IGT's Italian Software Architecture team, identified some features of the emerging blockchain, as enabling technologies to manage a ledger (or data store) that used cryptography and digital signatures to prove identity and authenticity (G. Ragnoni, E. Martire, F. Battini, *System and Methods for Blockchain-Based Digital Lottery Ticket Generation and Distribution* USPTO, Application Number 15/916,620 Filing date Mar 9 2018. https://patents.google.com/patent/US10931457B2). The idea was to design and implement a fully managed ledger database that provided a transparent, immutable and cryptographically verifiable platform for managing the creation of digital assets (e.g. game ticket) and the transfer of asset's ownership between users. Following the above idea, the team designed and implemented Transaction Certification Authority (TCA), an IGT platform that tracks each and every asset's transactions and maintains a complete and verifiable history of ownership change over time. The aim of this paper is to present TCA, its features and functioning mechanism.

Keywords Blockchain · Asset management · Digital game

1 Context and Concepts

The digitalization era is highlighting new opportunities to enhance customer experiences, providing new advanced, personalized, tailored and improved services for all actors involved into the Gaming Value-Chain [1].

Within this context, IGT built a digital gaming vision intending to unleash in-store digital experiences providing new digital journeys for customers within the Point of Sales. In this scope, IGT already provided several digital services to the customers enabling to get info, filling playslip, checking winnings through digital

G. Ragnoni (✉)
IGT, Rome, Italy
e-mail: gianluca.ragnoni@igt.com

A. Masone et al. (eds.), *Optimization and Data Science: Trends and Applications*, AIRO Springer Series 6, https://doi.org/10.1007/978-3-030-86286-2_14

touchpoints (e.g. mobile devices, self stations, etc.). In addition, mixed paper-digital services have already been provided enabling customers to participate to a digital lottery starting from a paper.

With the aim to have a full in-store digital experience the main challenge is related to provide the ability to manage digital tickets maintaining:

- The gaming model currently in place, where customers are not obliged to open game account to participate to the Lottery;
- The advanced security and anti-tampering measures integrated on the paper tickets.

Exploiting some features of the emerging Blockchain model [2, 3], IGT designed and developed a proprietary digital asset management solution called Transaction Certification Authority (TCA) for digital ticket management.

TCA is a fully managed ledger database that provides a transparent, immutable and cryptographically verifiable platform for managing the creation of digital assets and the transfer of assets' ownership between users. With TCA, the transaction's change history is immutable, it cannot be altered or deleted, using cryptography, and the client can easily verify that there have been no unintended modifications to asset's ownership.

The article is structured as follows: in Sect. 2 the TCA systems are described in terms of Scenario, Actors and Concept and Transactions available; in Sect. 3, the security requirements addressed by project referring to security standards are summarized; Sect. 4 is devoted to conclusions.

2 TCA

In this section the TCA systems are described in terms of Scenario, Actors and Concept and Transactions available.

The first subsection describes the main use case introducing a logical model to address the business requirements, the second subsection details the model of the project in terms of actors of the system, actions (or transaction) that each actor can do, rules of the actions, finally the type of transactions available in the system are explained using interaction diagrams.

2.1 Scenario

TCA is a platform designed to certify the ownership of arbitrary digital assets, regarding asset issuing and asset transferring between users, ensuring anonymity of the users, confidentiality, data integrity, non-repudiation, process transparency and reproducibility.

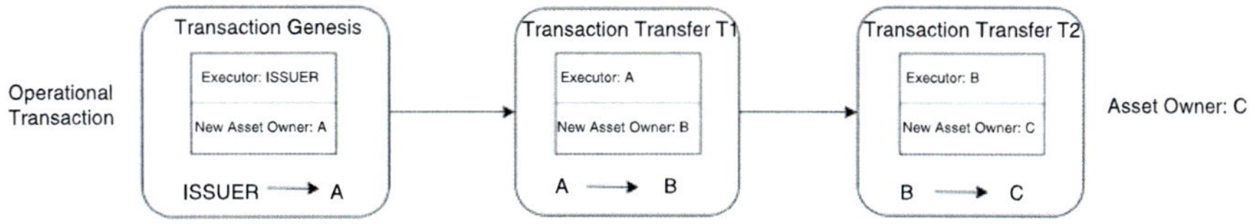

Fig. 1 Scenario A

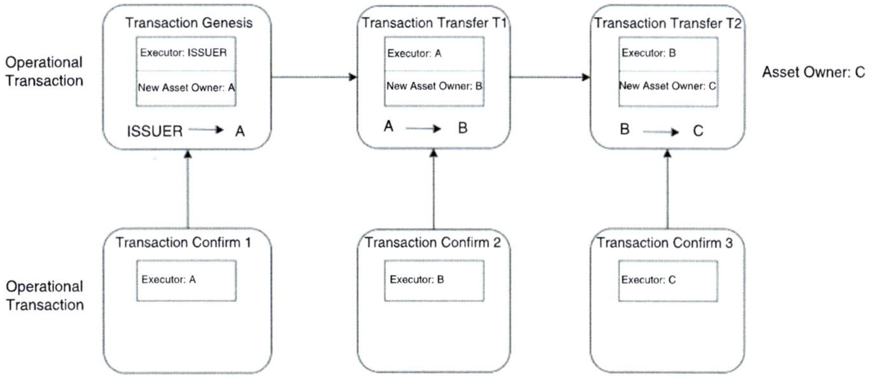

Fig. 2 Scenario B

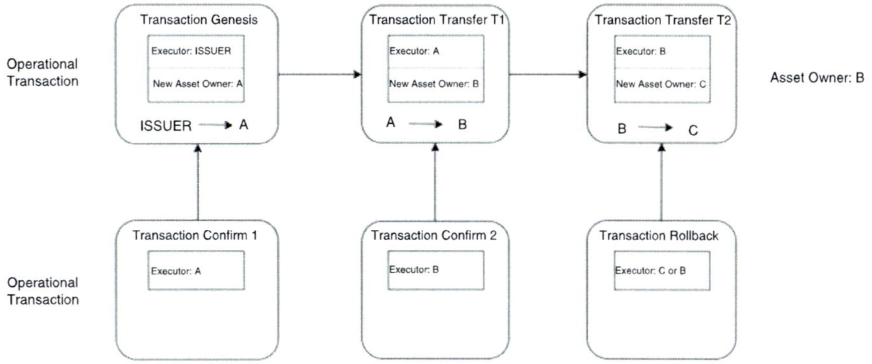

Fig. 3 Scenario C

By means of TCA a client with a specific clearance can register an asset on TCA; after registration users can transfer assets each other.

The TCA allows *implicit agreement* or *explicit agreement* when a new owner receives a digital asset: if the agreement is implicit (explained in the example in Fig. 1) when a digital asset is transferred to him, he automatically becomes the new owner, instead if the agreement is explicit (explained in the examples in Figs. 2 and 3) he may confirm or deny (rollback) the transfer.

2.2 Actors and Concepts

TCA adopts a model involving the following **actors**:

(i) The **Issuers**: the actors issuing digital assets;
(ii) **Clients**: the actors owning and transferring assets;
(iii) The Transaction Certification Authority (**TCA**);
(iv) One or more external Certification Authority (**CA**), used as a trusted third party external entity to guarantee immutability.

Each actor has a pair of asymmetric keys and it is identified by its public key (pubkey).

Issuer and customers interact through transactions.

2.3 Transactions

There are two types of **transaction**:

Operational Transaction Invoked by the subject issuing or transferring the asset; it allows to record the new ownership of the asset. Each transaction is the transfer of ownership of a digital asset (**DA**) from a public key called *SOURCE*, to another public key called *DESTINATION*. Each transaction is linked to the previous owner through the transaction hash.

- **Genesis Transaction**: this transaction allows to register an asset on the *TCA* and assigns its ownership to a client. It is always invoked by the *Issuer*.
- **Transfer Transaction**: allows to register the transfer of an asset from the current owner to a new owner.

Control Transaction If enabled, it is invoked by the client that receives an assets or, in case of rollback, by one any of the clients involved in the transfer, to confirm or deny the change of ownership. Each control transaction is linked to the operational transaction to be confirmed or rolled back.

- **Confirm Transaction**: this transaction allows the destination to confirm the willingness to receive the asset.
- **Rollback Transaction**: this transaction allows the destination to reject the asset or the source to abort the transfer if the destination has not confirmed it yet.

Each transaction is authorized by the TCA and the *SOURCE* using a digital signature.

Each transaction is recorded also in a block in a public transaction ledger (**PTL**) by TCA.

Each block is hashed and is linked to the previous block via hash value in the PTL.

Each block hash is also timestamped with a signature provided by CA. This feature guarantees the immutability via an external trusted third party entity.

Examples:

- **Scenario A**: in Fig. 1, control transaction disabled (implicit agreement). Only operational transaction. The final owner of the digital asset is C.
- **Scenario B**: in Fig. 2, control transaction enabled (explicit agreement), with confirm transactions. The final owner of the digital asset is C.
- **Scenario C**: in Fig. 3, control transaction enabled (explicit agreement), with confirm and rollback transactions. The final owner of the digital asset is B.

Figure 4 depicts a sequence diagram showing the message flow of Genesis Transaction with implicit non-repudiation. Genesis Transaction has three actors: a customer (Customer1) requiring an asset generation from an authorized issuer (Issuer) that generates the asset, the TCA as a digital notary to guarantee issuing, transfer and ownership of the digital asset.

In message #1 Customer1 requires asset issuing to the Issuer, specifying his public key as the destination for the asset.

In message #2 Issuer creates a digital asset (DA) and creates a message (MSG) contains DA, the origin of DA (or the source) inserting his public key and the destination of the asset using the public key received from the Customer1.

In message #3 Issuer signs MSG with his private key (DS Issuer).

In message #4 Issuer sends MSG and DS Issuer to the TCA.

In message #5 TCA does a set of checks as for example if the Issuer is authorized to issue the asset, if asset is new and it is well formed syntactically etc . . . If checks are positive, TCA signs MSG (DS TCA), creates a new transaction with MSG, DS Issuer and DS TCA and saves the transaction in the Public Transaction Ledger (PTL).

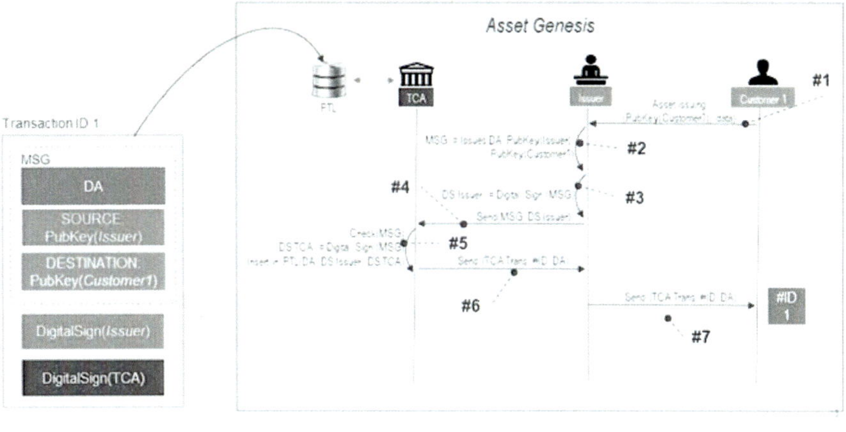

Fig. 4 Genesis transaction

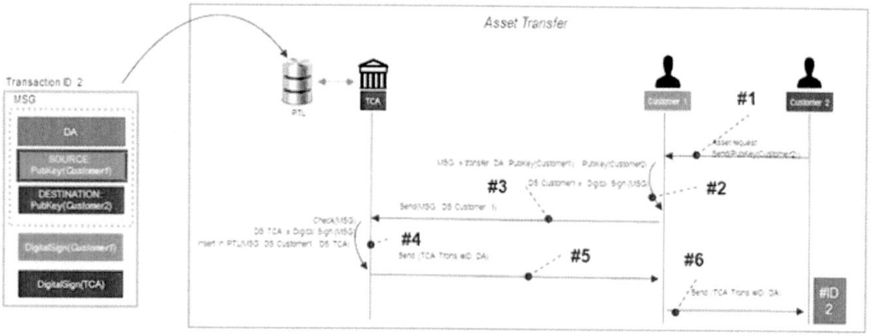

Fig. 5 Transfer transaction

In messages #6, #7 transaction is sent back to Issuer and to the Customer1. At the end of this process, Customer1 is the owner of the asset.

Looking at the Transaction ID1 to the left of the diagram, we can see the message (MSG) containing the asset (DA), the issuer of the asset (or SOURCE) represented by the public key of the Issuer, the owner of the asset (or DESTINATION) represented by the public key of the Customer1. Then the message signatures made by the Issuer and by the TCA.

Figure 5 depicts a sequence diagram showing the messages flow of Transfer Transaction with implicit non-repudiation. A typical Transfer Transaction has three actors: a customer (Customer2) requiring the transfer of an asset from another customer (Customer1) that owns the asset and the TCA that acts as a digital notary to guarantee correctness of transfer and new ownership of the digital asset. In the Transfer Transaction the current owner of the asset (SOURCE), has the right to transfer the asset to someone else (DESTINATION).

In message #1 Customer2 requires an asset transfer to Customer1 specifying his public key as the new destination of the asset. In arrow #2 Customer1 creates a message (MSG) contains DA, the origin of DA (or the source) inserting his public key and the destination of the asset using the public key received from the Customer2. Customer1 signs MSG with his private key (DS Customer1).

In message #3 Customer1 send MSG and DS Customer1 to TCA.

In message #4 TCA does a set of checks as for example if the Customer1 is the current owner of asset and has the right to transfer. If true, TCA digital sign MSG (DS TCA), creates a new transaction with MSG, DS Customer1 and DS TCA and saves the transaction in the Public Transaction Ledger (PTL).

In messages #5, #6 transaction is sent back to Customer1 and to the Customer2. At the end of this process, Customer2 is the owner of the asset.

Looking at the Transaction ID2 to the left of the diagram, we can see the message (MSG) containing the asset (DA), the previous owner of the asset (or SOURCE) represents by the public key of the Customer1, the current owner of the asset (or DESTINATION) represents by the public key of the Customer2, the message signatures made by the Customer1 and by the TCA.

3 Security Features

According to the standard **ISO/IEC 27000:2018** [4], information security is the protection of confidentiality, integrity, availability (often denoted by acronym CIA); the same standard also includes other requirements, as authenticity, accountability, non-repudiation, and reliability. In the well-known glossary CNSS [5], data security is described according to the CIA requirements: "*The protection of information and information systems from unauthorized access, use, disclosure, disruption, modification, or destruction in order to provide confidentiality, integrity, and availability*".

The definition in the ISACA [6] glossary is similar: "*Ensures that only authorized users (confidentiality) have access to accurate and complete information (integrity) when required (availability).*"

In the framework of the TCA project, among the ISO/IEC 27000:2018 requirements, we select as relevant, the following items to implement:

- Confidentiality: information is not made available or disclosed to unauthorized individuals, entities, or processes;
- Integrity: information is complete and correct, in particular, information cannot be modified in an unauthorized or undetected manner;
- Authenticity: each entity involved in some operation should provide a proof that it is what it claims to be;
- Non-Repudiation: each entity involved in some operation cannot deny having executed its own actions.

Other requirements are not managed directly, since they are implied by the previous ones (accountability is implied by integrity, authenticity and non-repudiation), or they are out of the scope of this study (availability, reliability).

4 Conclusion

In the gaming context, TCA platform can enable ticket receipt dematerialization inside the retail store with a seamless integration with current ecosystem providing a "Phygital" (physical and digital) experience for the user that can have a digital version of ticket receipt (the asset) in their mobile device, with the same security features of the physical one.

Moreover TCA, can play the role of "layer 2 infrastructure" integrating itself with a public blockchain infrastructure. In fact, since 2017, public blockchain infrastructures have been evolved and nowadays, although there are a lot of blockchain available, capable to manage directly arbitrary digital assets, according to experts and community thoughts [7], specific domain applications will be necessary on top of blockchain, in order to manage specific digital assets in a scalable way. Playing this role, TCA could be a specific domain application that manages digital tickets for gaming companies that want to enable ticket receipt dematerialization. In this scenario, public blockchain infrastructure can replace the Certification Authority saving only block's fingerprint of transactions managed by TCA.

TCA model has been patented [8] in US with number US10931457B2 and is already integrated in the Italian Digital National Lottery.

References

1. The World Lottery Summit 2018—Jean Jorgensen Merit Award for Innovation—"Lottomatica Lottery Ticket Digitalization Based on Blockchain Model"
2. Bitcoin: A Peer-to-Peer Electronic Cash System—Satoshi Nakamoto. https://bitcoin.org/bitcoin.pdf
3. Blockchain Series (MOOC)—University at Buffalo School of Engineering and Applied Sciences. https://www.coursera.org/learn/blockchain-basics
4. ISO/IEC 27000:2018. ISO/IEC 27000:2018. https://www.iso.org/standard/73906.html (2018)
5. Committee on National Security Systems. National Information Assurance (IA) Glossary, CNSS Instruction No. 4009, 26 Apr 2010. https://www.hsdl.org/?view&did=7447
6. ISACA. Information Systems Audit and Control Association. http://www.isaca.org/Knowledge-Center/Documents/Glossary/glossary.pdf (2021)
7. 102 blockchain leaders share their insights into the use of blockchain both now and into the future. Taken from https://zage.io/report/pdf/blockchain-102.pdf
8. Ragnoni, G., Martire, E., Battini, F.: Systems and methods for blockchain-based digital lottery ticket generation and distribution. USPTO, Application Number 15/916,620—Filing date 9 Mar 2018. https://patents.google.com/patent/US10931457B2